PHOTO MANUAL and DISSECTION GUIDE of the SHARK

Fred Bohensky, Ph.D.
The College of Staten Island
City University of New York

Photographs by
Joe Rickard

AVERY PUBLISHING GROUP INC.
Wayne, New Jersey

Avery's Anatomy Series

Photo Manual and Dissection Guide of the Cat
Fred Bohensky

Photo Manual and Dissection Guide of the Fetal Pig
Fred Bohensky

Photo Manual and Dissection Guide of the Shark
Fred Bohensky

**Understanding the Human Form: A Practical Study
& Workbook for Human Surface Anatomy**
Frank D. Shapiro
Arnold Burchess
Harriet E. Phillips

Avery illustrations on pages 3, 5, 30, 44, 58, 76, 90, 96, 102 and 116 drawn by Scott J. Shevins.

Copyright © 1981 by Fred Bohensky

ISBN 0-89529-140-1

All rights reserved. No part of this publication may be reproduced, stored in a retrieval system, or transmitted, in any form or by any means, electronic, mechanical, photocopying, recording, or otherwise, without the prior written permission of the copyright owner.

Printed in the United States of America

CONTENTS

List of Photographs and Illustrations...iv

Preface..vii

Introduction: The Shark and Man..1

Anatomical Terminology...3

External Features..5

General Dissection Hints..14

The Skeletal System..16
   Self-Quiz I...29

The Muscular System..31
   Self-Quiz II..43

The Digestive and Respiratory Systems..45
   Self-Quiz III...57

The Circulatory System...59
   Self-Quiz IV..75

The Urogenital System..77
   Self-Quiz V...89

The Nervous System: The Brain and Spinal Cord..................................91
   Self-Quiz VI...101

The Nervous System: The Sense Organs..103
   Self-Quiz VII..115

Shark Dissection Check List...117

LIST OF PHOTOGRAPHS AND ILLUSTRATIONS

Anatomical Terminology
    Shark, Directional Terms, *Illustration*, 3
    Man, Directional Terms, *Illustration*, 4

External Features
    Skin Structure, *Illustration*, 5
    Shark, External Features, 9
    Head, Lateral View, 10
    Head, Ventral View, 11
    Pelvic Region, Male, 12
    Pelvic Region, Female, 13

General Dissection Hints
    Dissection Tools, *Illustration*, 15

The Skeleton System
    The Skeleton, Ventral View, 21
    Skull, Dorsal View, 22
    Skull, Ventral View, 23
    Skull, Lateral View, 24
    Caudal Vertebra, Cross Section, 25
    Pectoral Girdle and Fins, Ventral View, 26
    Pelvic Girdle and Fins, Male, Ventral View, 27
    Pelvic Girdle and Fins, Female, Ventral View, 28
    The Skeleton, Ventral View, *Illustration*, 30

The Muscular System
    Anterior Musculature, Ventral View, 36
    Anterior Musculature (Close-up), Ventral View, 37
    Anterior Musculature, Deeper Muscles, Ventral View, 39
    Anterior Musculature, Lateral View, 39
    Posterior Musculature, Lateral View, 40
    Posterior Musculature, Cross Section, 41
    Posterior Musculature, Stereoscopic View, 42
    Anterior Musculature, Lateral View, *Illustration*, 44

The Digestive and Respiratory Systems
    The Viscera, Ventral View, 49
    The Viscera (Close-up), Ventral View, 50
    The Dorsal Viscera, 51
    The Stomach, 52
    The Valvular Intestine, 53
    The Oral Cavity and Pharynx, 54
    The Gills (Close-up), 55
    Gill Surface, 56
    The Viscera, Ventral View, *Illustration*, 58

The Circulatory System
    The Heart, Ventral View, 67
    The Branchial Arteries, 68
    The Branchial Arteries (Close-up), 69
    Major Blood Vessel of Trunk, 70
    Major Blood Vessels of Trunk (Close-up), 71
    Systemic Veins, 72
    Systemic Veins (Close-up), 73
    Major Blood Vessels of the Trunk, *Illustration*, 76

The Urogenital System
    The Female Urogenital System, 81
    The Female Urogenital System (Close-up), 82
    Embryos Within Uterus, 83
    Embryos Removed from Uterus, 84
    Mature Embryo in Uterus, 85
    The Male Urogenital System, 86
    Male Urogenital System (Cloaca Exposed), 87
    Male Urogenital System (Close-up), 88
    The Female Urogenital System, *Illustration*, 90

The Nervous System: The Brain and Spinal Cord
    Sensory and Motor Impulses Diagram, 96
    The Brain and Cranial Nerves, Dorsal View, 97
    The Brain and Cranial Nerves (Close-up), Dorsal View, 98
    The Brain, Sagittal Section, 99
    The Vagus and Hypobranchial Nerves, Dorsal View, 100
    The Brain and Cranial Nerves, Dorsal View, *Illustration*, 102

The Nervous System: The Sense Organs
    The Membranous Labyrinth, Dorso-Lateral View, 107
    The Eye, Dorso-Lateral View, 108
    The Eye, Dorsal View, 109
    The Eye, Sagittal Section, 110
    The Left Orbit, Lateral View, 111
    The Left Eye, Medial View, 112
    The Olfactory Sac, Cross Section, 113
    The Eye, Dorsal View, *Illustration*, 116

DEDICATION

To my wife Esther, who gave up yet another
summer vacation to type, proofread, and work along
with me to see this book to its completion.

PREFACE

The spiny dogfish shark continues to be a favorite dissection specimen in courses of Zoology, Comparative Anatomy, and Vertebrate Zoology. It is generally the first animal to which the student in such classes is exposed.

It presents fine examples of structural patterns of the primitive fish, while a thorough study reveals that the shark holds the key to the higher vertebrates, as well as to man himself. It is with fascination that structures such as those of the circulatory, excretory, and nervous systems are traced from their more primitive manifestations in the shark to their more complex expressions in amphibians, reptiles, birds, and mammals.

The relative large size of the shark makes dissection that much easier. It permits a study of muscles, blood vessels, and nerves which cannot be pursued with smaller, immature animals. Unlike the fetal pig, fully formed adult structures are observed.

Even the second year biology student is often overwhelmed when, at the start of the semester, he or she is handed a dissection specimen and instructed to follow the procedures outlined in the lab manual. He will encounter difficulty in relating the structures seen in the specimen before him with the diagrams in his lab manual. These are often poorly drawn, small, or inaccurately labeled. To the untrained eye, nerves, blood vessels, and smaller structures look very different from the figures as depicted. They are often idealized versions of "perfect" dissections.

The Photo Manual and Dissection Guide of the Shark presents full-page photographs revealing the various structures as they will appear to the student. Some are as big as the actual specimens, some are even larger. These have been carefully selected and labeled. Where doubt might exist as to the limits of an organ or a structure, these have been outlined on the photo with dotted lines, or their names have been printed directly on the organ. Some of the photos show the same areas in different views or in close-up. This will help students to pinpoint structures.

The sharks used in these model dissections were young adults, about 2 to 2½ feet long. Arteries and veins were injected with red and blue latex dyes respectively, the hepatic portal system with yellow dye. The head, tail, pectoral, or pelvic fins have been included in many of the photographs to facilitate identification.

A Self-Quiz for students is found at the end of each unit. These consist of short answer type questions, definitions, and the labeling of review diagrams. These worksheet pages may be removed from the book and submitted to the instructor for correction.

A *Dissection Check List* is provided at the close of this manual. It includes all of the structures described in the text. At the outset of the study of each physiological or anatomical system, the instructor may indicate to the students which organs they are responsible for identifying and locating. Each instructor will assign those items from the Check List deemed most important, thus providing greater flexibility in the syllabus.

One structure may have several names as used by different authors. This is particularly true for many muscles and blood vessels. In such cases, the most commonly used name or the one most descriptive will be indicated; the alternate name, or names, will appear in parentheses.

The author wishes to acknowledge the help extended by the Media Production Center of The College of Staten Island. He is particularly grateful for the assistance of Mr. Joseph Rickard, the photographer, whose fine lenswork is seen throughout this manual.

Fred Bohensky
Staten Island, N.Y.

INTRODUCTION: THE SHARK AND MAN

*"Who can open the jaws of his face?
His teeth are terrible round about."*
Job, 41:6

The very mention of the word shark has from ancient times instilled within man an almost irrational fear. Yet, at the same time, the cunning, strength, tenacity, and grace of movement of the animal have never ceased to fascinate him.

Sharks have in the past been studied by ichthyologists alone. Today, with the popularity of such films as *Jaws I* and *Jaws II* and the publication of numerous paperbacks on the subject, even the person without a science background is aware of the more dramatic aspects of shark behavior.

The present day popular interest in sharks can be traced to the wartime experiences of the armed forces. The sinking of troop transport ships and the shooting down of aircraft over the open seas often ended in carnage resulting from shark attacks upon those who had survived the guns and torpedoes.

This led to the founding of the *Shark Research Panel* as a joint venture of the U.S. Navy and the American Institute of Biological Science (AIBS). Its aims include:

(1) To keep an up-to-date record of shark attacks throughout the world, and to compile annual statistics of these accidents, published as a list with an analysis of the attack and the conditions under which it occurred.

(2) To promote and to follow up all basic research by coordinating studies carried out in all parts of the world on systematics, migrations, general biology, anatomy, and physiology of sharks.

(3) To study the methods and results of shark repellent investigations.

SHARKS AND FISH

Biologically, sharks are fish belonging to the phylum *Chordata* and the subphylum *Vertebrata*. However, sharks and their relatives, the rays and skates, are unique amongst fish in that their skeletons are made entirely of cartilage, not bone. This places them in the class *Chondrichthyes*, subclass *Elasmobranchii*.

The bony fish, the *Osteichthyes*, possess a gas-filled *swim bladder* by means of which they can regulate their buoyancy allowing the fish to "float" at various depths under water. Sharks have no swim bladders. They are somewhat heavier than the water they displace. Thus, once a shark ceases to move, it sinks. Coastal species rest on the sea floor in shallow water. However, the sharks of the deeper oceans must continue moving from the moment of birth to the moment of death! If they were to stop swimming, they would sink and be crushed by the pressure of the deep below.

Regulation of *osmotic pressure* in marine sharks differs from that of their bony relatives. They retain a high concentration of urea and other solutes in their body fluids, a concentration of salts higher than that in the surrounding sea water. There is therefore no need for sharks to drink.

Fertilization is *internal*, and most shark "pups" hatch internally, to continue their development within the *uterus* of the mother. After a period of gestation (up to two years in the spiny dogfish, *Squalus acanthias*, the longest of any vertebrate!) they are born alive as a smaller version of the adult. This method of reproduction is called *ovoviviparous*. The number of "pups" in a litter varies from two in some species to sixty in others. Some sharks are *oviparous*, laying large eggs enclosed in shells, or egg-cases, consisting of a hornlike material. They are usually flat and quadrangular shaped with long tendrils which serve to anchor the eggs to seaweed or other objects.

While there are close to 20,000 living species of fish, only about 300 of these are sharks. They are divided into nineteen families, with five families making up 75 per cent of the known species.

Sharks range in size from a species only six inches long when mature to the 35-foot basking shark and to the largest of all fish, the *whale-shark*, reaching 50 feet in length and weighing over ten tons. Contrary to popular belief, these two largest of sharks are quite inoffensive beasts, deriving most of their nourishment from minute planktonic animals (those which float in the upper layers of the sea).

SPINY DOGFISH SHARK

The spiny dogfish, genus and species names *Squalus acanthias,* of the family Squalidae, is our dissection specimen. The species name "acanthias" calls attention to the animal's mildly poisonous *spines,* one in front of each dorsal fin. It is a relatively small shark, attaining about 3½ feet in length and weighing about 15 pounds. The absence of an anal fin is characteristic of the entire family.

It is distributed worldwide, from the temperate to the subpolar regions, from the shallow waters of the seashore to depths of 100 fathoms (600 feet).

They are voracious eaters, feeding on fish, crustaceans, squid, gastopods, jelly fish, and even red and brown algae. The spiny dogfish, as most other sharks, is *omnivorous,* devouring both plant and animal matter.

It is an abundant species. On this side of the Atlantic it is infamous for its disruptive activities to fishing operations. It is destructive of fishing gear; hook and line, nets are bitten and their catch devoured and freed. This results in a high animal loss to the fishing industry. Except as laboratory specimens, no economic use has been found for them.

In northern Europe and the British Isles, the spiny dogfish shark is similarly destructive but is there counted as a food fish species. Known as the *spurdog* or *piked dogfish,* a great deal of it is served in fish and chips shops.

Tagging studies indicate that the animal has a life span of 25-30 years. The male reaches sexual maturity at the age of 11, the female at 19. Litters are small, generally four to seven pups. The excessively long gestation period, two years, longest of any other vertebrate, has already been mentioned.

In the laboratory they serve two functions. Firstly, they serve as dissection specimens, illustrating many primitive vertebrate features, and thus useful in tracing these features through the higher vertebrates to man. Secondly, their relatively small size is a convenience in housing and caring for live specimens in a broad range of physiological research situations, such as human cardiology and immunology.

Man, *Homo sapiens*, belongs to the class *Mammalia*, whose members:

— Possess milk producing glands in the female (mammary glands) for nursing the young;

— Have skin covered with hair or fur.

The dogfish shark is a comparatively large dissection specimen. Its muscles and internal organs are clearly visible. Its nerves and blood vessels are readily traced.

The photographs and diagrams accompanying each chapter are for your use as learning tools. Identify and learn the names of the structures of your specimen by repeated referrals to the photographs.

ANATOMICAL TERMINOLOGY

Some basic biological terminology should be studied at this time. Familiarize yourself with the following words and learn to use them in referring to the location of the body parts of your specimen.

DIRECTIONS OR POSITIONS

Anterior (Cranial)   — toward the head
Posterior (Caudal)   — toward the tail
Dorsal (Superior)   — toward the backbone
Ventral (Inferior)   — toward the belly
Lateral   — toward the side
Medial   — toward the midline
Proximal   — lying near the point of reference
Distal   — lying further from the point of reference

Note: The terms in parentheses are synonymous when referring to a fish-like animal such as the shark or to a quadruped such as a cat. In man these terms have different meanings (see diagram at the end of this section).

PLANES OR SECTIONS THROUGH THE BODY

Transverse (Cross Section) — perpendicular to the long axis of the body
Sagittal — a longitudinal section separating the body into right and left sides
Frontal (Coronal) — a longitudinal section dividing the specimen into dorsal and ventral parts

DIRECTIONAL TERMS for the Shark and Man

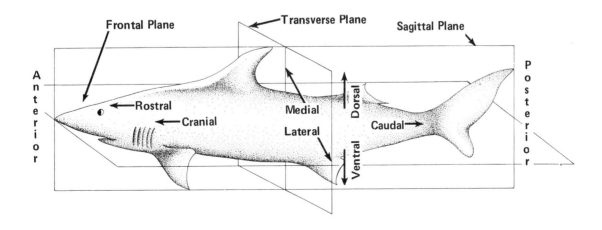

In man, the *anterior* and *ventral* surfaces are identical; both terms refer to a person's front or belly side. Similary, the terms *posterior* and *dorsal* are identical, referring to a person's back surface, the area near the spinal cord.

Other terms indicating position or direction will appear in the text. For example, the terms *superficial* (or external) and *deep* (or internal) will be used when describing muscles. The terms *pelvic* and *pectoral* will describe different fins of the shark.

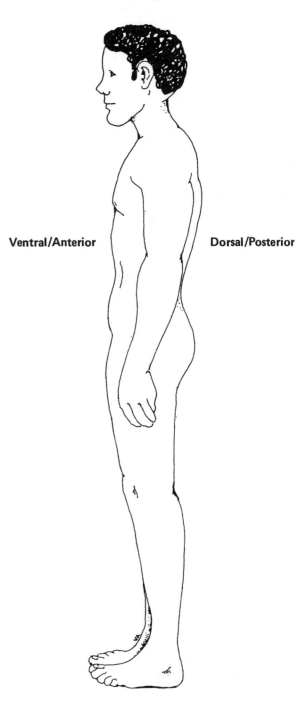

EXTERNAL FEATURES

Specimens of dogfish shark generally come individually packed in large plastic bags together with about a pint of preservative solution. Prepare a large dissection pan and some paper towels. Open the bag and carefully remove the animal onto the pan. Do not discard the preservative. It should remain in the bag during the entire dissection. Examine your specimen.

OVERALL DESCRIPTION

The shark is gracefully elongated and streamlined. The body shape is known as *fusiform*, built for swimming in the sea with least possible resistance.

The student should now refer to the first photo.

The body is divided into three readily identifiable areas:

The head (cranial) — from the pointed snout-like *rostrum* to the *pectoral fins*. This includes the *gill* region.

The trunk — from the *pectoral fins* to the *pelvic fins*.

The tail (caudal) — from the *pelvic fins* to the end of the *caudal fin*.

THE SKIN

Run your hand over the body of the shark from head to tail and feel its smooth texture. Now, run your hand in the opposite direction and you will detect a rough, sandpaper-like texture. As a matter of fact, shark skin has been used as an abrasive in the manufacture of furniture for hundreds of years. It was also used as a covering for sword handles and tools to prevent them from slipping from one's hand.

The entire skin of the shark is covered by minute, sharp, tack-like *placoid scales* embedded in the skin pointing caudally. These scales differ considerably from the oval overlapping transparent scales of most bony fish. They are modifications of teeth; thus their name, *dermal denticles*. Their structure and mode of development are similar to the teeth of higher vertebrates and man. See diagram below.

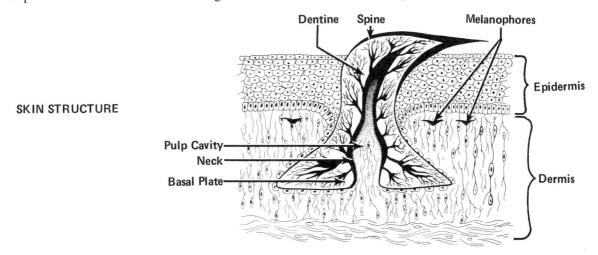

SKIN STRUCTURE

Like true teeth, the placoid scales have a base of *dentine* which contains a *pulp cavity* filled with connective tissue. Both scales and teeth have a spinous process covered by *enamel* which protrudes through the skin.

The shark's body is colored dark gray above and much lighter, almost white, below. This distribution of pigment is referred to as *counter shading* and is common amongst aquatic vertebrates. It tends to neutralize the effects of natural lights, which, coming from above, highlights the back and casts a shadow on the underside. It tends to make the animal less conspicuous.

Extending laterally, along the sides of the body, somewhat nearer to the dorsal than to ventral surface, look for a narrow, light-colored horizontal stripe. Observe carefully along this line with a magnifying glass or low power dissection microscope and note the pores along its length. This is part of the *lateral line system*. Below the skin, nerve receptors called *neuromasts* run along a *lateral line canal* with pores opening to the surface. They carry impulses to the central nervous system. These receptors, found only in fish and some aquatic amphibians, are sensitive to the mechanical movement of water, to disturbances in the water, and to sudden changes of pressure. They warn the shark of vibrations and movements even in murky water, where visibility is reduced.

In the area of the head the lateral line canal branches to form several communicating canals. These will be described in the chapter on the Nervous System, page 106.

Note patches of pores upon the head in the areas of the eyes, snout, and nostrils. These are the openings of the *ampullae of Lorenzini*, sense organs which are sensitive to changes in temperature, water pressure, electrical fields, and salinity. Press firmly upon the skin near the nares (nostrils). Note the jelly-like material you have squeezed out of the pores. These will be examined further in the chapter on the Nervous System, page 106.

THE HEAD

Examine the head of the shark and note each of the following:

ROSTRUM — This is the pointed snout at the anterior end. This streamlined tapered tip at the anterior end helps overcome water resistance in swimming. See the first three photos at the end of this chapter.

NARES — These are the openings for the external nostrils. They are located on the underside (ventral surface) of the rostrum anterior to the jaws. Water is drawn into the nares to moisten the sensory cells of the olfactory sac. A *nasal flap* can be seen clearly in the photo of the ventral surface of the head, page 11, which separates the *incurrent aperture* from the *excurrent aperture*. Water passes into and out of the olfactory sac, permitting the shark to detect the odors of the water. The ability of the sharks to detect blood and injured flesh at great distances from their source is legendary and is a major attractant and subsequent cause of shark attacks.

JAWS — The opening to the mouth of sharks is always on the underside. The great powers of the shark's jaws have been retold by mariners for generations. Recently a testing device, the *gnathodynamometer*, was used to measure the force exerted by the jaws of a typical eight-foot shark. It was an extraordinary eighteen tons per square inch!

Examine the *teeth*. They are sharp and pointed. They are formed of the same material and develop similarly to the smaller *placoid scales* distributed over the shark's entire body. Besides the visible teeth, several rows of flattened teeth lie behind the upright set ready to replace them when worn out or lost. It has been estimated that the great white shark has about 400 teeth.

EYES — These are very prominent in sharks, and as our later dissection and photo will reveal, see pages 104 and 108, they are very similar to the eyes of man.

A transparent *cornea* covers and protects the eye.

A darkly pigmented *iris* can be seen below the cornea. Its contraction and relaxation adjusts the amount of light entering. In its center is the opening to the interior of the eyeball, the *pupil*.

Upper and lower eyelids protect the eye. Just inside the lower lid, a membrane may be seen, the *conjunctiva*. It extends over the surface of the eye to cover the cornea.

SPIRACLES — These large openings posterior and dorsal to the eyes are actually reduced first gill slits in the jawed fish. A *pseudobranch,* false gill, is a reduced first gill which may be seen within the spiracle. A fold of tissue, the *spiracular valve,* permits the opening and closing of the external spiracular pore. The spiracle serves as an incurrent water passageway leading into the mouth. Thus water can be brought in for respiration even when the shark's mouth is closed or when he is feeding.

GILL SLITS — Most sharks have five external gill slits. They are located laterally, posterior to the mouth, in front of the pectoral fins. Water taken in by the mouth is passed over the internal gills, oxygen is removed and carbon dioxide excreted. The water is then forced out to the external environment by way of the gill slits. The structure of the gills, their cartilaginous support and blood supply will be discussed in later chapters.

ENDOLYMPHATIC PORES — Look at the top of the head between the spiracles with a hand lens. You will see a pair of tiny *endolymphatic pores,* one on each side of the midline. They are continuations of the *endolymphatic ducts* which lead into the *inner ear* which, in turn, serves primarily as an organ of equilibrium.

THE FINS

The spiny dogfish shark possesses two single dorsal fins, a caudal fin, and two pairs of ventral fins.

DORSAL FINS — The anterior dorsal fin is larger than the posterior dorsal fin. When sharks are seen near the surface of the water, the telltale sign is the triangular anterior dorsal fin projecting ominously above the surface of the water.

A feature peculiar to our specimen, the spiny dogfish, is the presence of two *spines,* one immediately anterior to each dorsal fin. When captured, these sharks will arch their backs and attempt to pierce their captor with these long sharp spines. Besides the puncture wounds these can inflict, the spines also carry a poison secreted by glands at their base. The structures and origins of these spines are similar to those of the tiny dermal placoid scales and teeth.

CAUDAL FIN (Tail Fin) — This fin is divided into two lobes; the larger *dorsal lobe,* and smaller *ventral lobe.* Note that the tapering body axis passes upwards into the dorsal lobe. This type of tail is known as a *heterocercal* tail, as opposed to the single-lobed, fan-shaped symmetrical tail of the bony fish known as a *homocercal* tail.

PECTORAL FINS — The asymmetry of the shark's tail fin creates a problem. As the tail is moved back and forth, the larger dorsal lobe causes the shark to be propelled forward and downward in the water. To offset the downward tendency, the paired pectoral fins act to deflect water downward and thus provide the lift needed at the crucial end to keep the shark moving in a horizontal direction.

PELVIC FINS — These paired ventral fins are located on either side of the cloacal aperture. They are different in males and females; see photos, pages 12 and 13. Those of the female are undifferentiated while those of the male are specialized for use in the transfer of sperm to the female during copulation or mating.

THE CLOACA — This name is given to the chamber on the ventral surface between the pelvic fins. It receives the products of the intestine, the urinary and the genital ducts. The name, meaning sewer, seems quite appropriate. A closer look within the cloaca will reveal the *urinary papilla.* Also visible, especially in mature female specimens, are the *abdominal pores.* These will be discussed later; see page 47.

CLASPERS — Males have stout, grooved copulatory organs called *claspers* on the medial side of their pelvic fins. Fertilization in the dogfish shark is internal. During copulation, one of the claspers is inserted into the oviduct orifice of the female. The sperm proceed from the cloaca of the male along the groove on the dorsal surface of the clasper toward the female.

Associated with the claspers of the male are accessory structures such as the *siphons* and in some specimens *lateral spines* and *ventral hooks* may be present near the end of the claspers. The relationship of the pelvic fins to the cloacal openings as well as some of the accessory structures in males and females are better seen in the photos in the chapter on The Reproductive System, pages 82 and 86.

EXTERNAL FEATURES

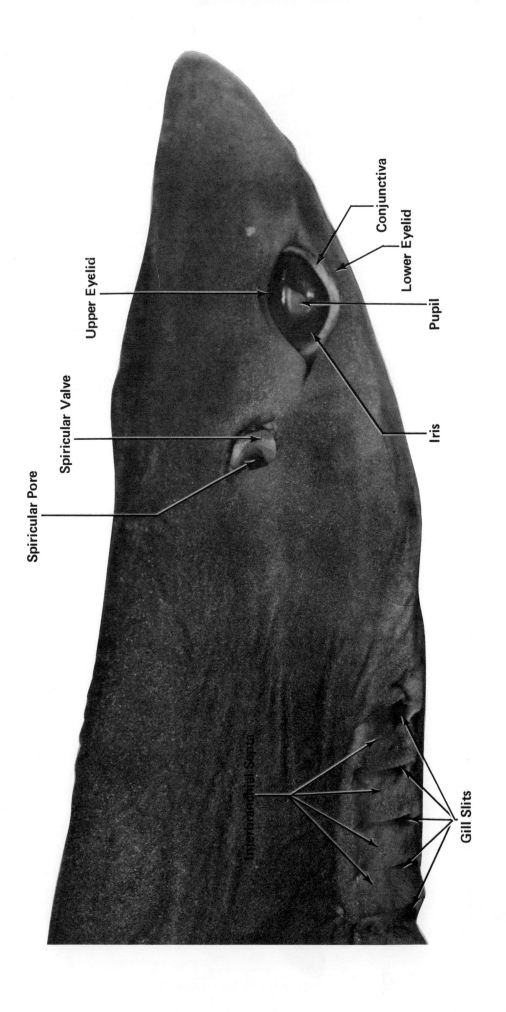

HEAD – LATERAL VIEW

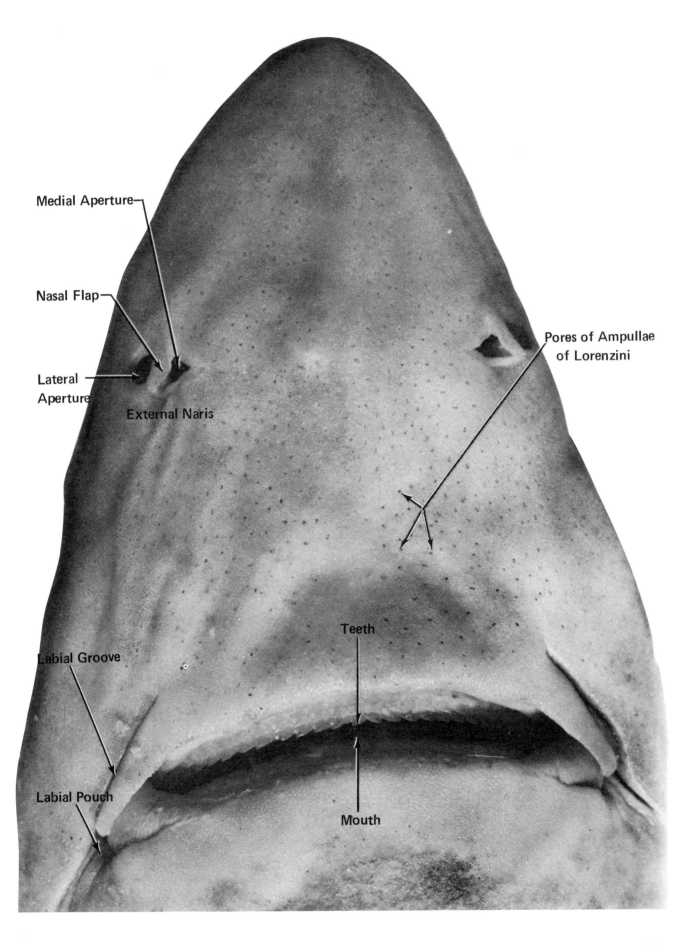

HEAD – VENTRAL VIEW

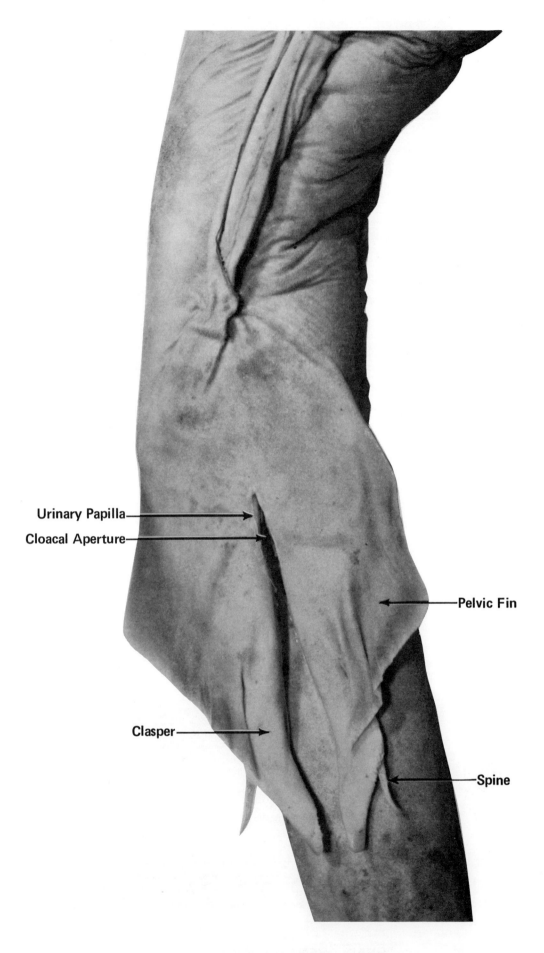

PELVIC REGION — MALE

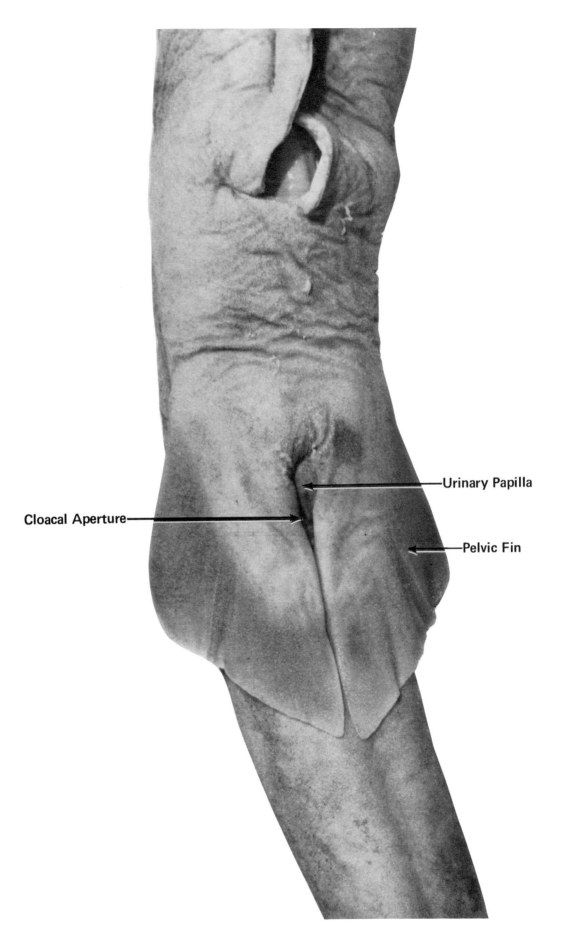

PELVIC REGION — FEMALE

GENERAL DISSECTION HINTS

The term "dissection" means more than merely cutting your specimen apart. It is a refined method of seeking, exposing, identifying, and studying internal anatomy. It helps to bring into view structures not readily seen.

The dogfish shark is generally not the first organism dissected by students. Most likely a frog or fetal pig has already been studied in previous courses. Based upon the earlier experiences, more is expected of the student, a better dissection. You will note that more difficult tasks are presented, such as, exposing the delicate semicircular canals, embedded entirely in cartilage. A similarly difficult task is the exposure of the brain and cranial nerves of the shark.

A cartilaginous skeleton as found in the shark offers certain advantages. It permits penetrating a tissue that is ordinarily hard and bony. This requires a certain amount of skill, for cutting into the cartilage can damage nerves and other structures irreparably. The technique of slicing thin chips while holding the scalpel horizontally must be practiced. A slip of the blade may undo hours of careful work.

Use your *scalpel* sparingly. In the hands of a novice a scalpel can do irreparable damage to your specimen. Blood vessels and nerves may be cut, organs removed, delicate structures destroyed without realizing the extent of the damage caused. Improper initial dissection will render the later study of parts very difficult.

The rough skin of the shark and the repeated cutting of cartilage will dull your scalpel and necessitate changing your scalpel blade several times during the dissection.

Rely more heavily upon your *dissecting needles*, your *blunt probe, flexible probe*, and even your *fingers* (several common dissecting tools are illustrated on the following page). They are especially helpful in separating muscles, in tracing blood vessels and nerves, and in clearing away connective tissue that binds structures to one another.

When using your *scissors*, advance with the rounded, blunt end, not the sharp, pointed end. Your *forceps* should be strong, able to hold on to thick muscle, yet fine enough to grasp narrow nerves. It is advisable to have more than one type of forceps. Move organs aside with your fingers or with a blunt probe.

Your animals are preserved in a solution which may irritate your hands and eyes. Your laboratory should be well ventilated to prevent the buildup of fumes. A lab coat or apron will protect your outer clothing. It is also suggested that you apply lanolin or petroleum jelly (vaseline) to your hands at the start of each dissection or wear thin plastic or rubber gloves. Line your dissection pan with paper towels in order to absorb excess fluids, as a storage for structures removed, and to facilitate cleaning up at the close of the session.

Observe the dissections of other students in the class. Often a better preserved, a better injected, or a larger specimen may reveal structures not seen in your shark. This is especially true in the study of the urogenital system. If your animal is a male, observe the reproductive structures of a female specimen and vice versa. You are responsible for learning the reproductive structure of both male and female sharks.

At the end of each session the sharks are returned to the plastic bag. Twist the top of the bag and close tightly with a rubber band. These procedures will prevent the drying out of your specimen between dissections. Remove the paper towels lining the dissection pan, together with any structures removed, and discard.

In order to preserve further the softness and texture of the exposed shark musculature and viscera, apply the following solution with a one-inch paint brush at the close of each session:

Carbolic Acid (Phenol) crystals	30 grams
Glycerin	250 ml.
Water	1000 ml.

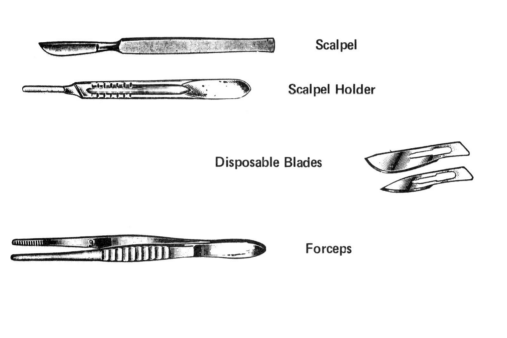

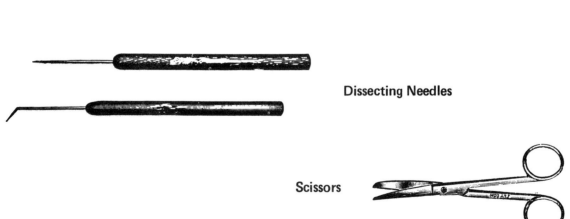

DISSECTION TOOLS

THE SKELETAL SYSTEM

INTRODUCTION

The internal *endoskeleton* of vertebrates provides support and protection for the softer body parts. It is the fundamental system upon which the body is built. A paleontologist can reconstruct an entire organism from a study of its bones alone.

Two types of skeletal bones are recognized in the higher vertebrates. The first is *endochondral bone* where solid bone has replaced earlier embryological cartilage. These are exemplified by our arm and leg bones. Others, known as *dermal* or *membrane bones,* have no cartilaginous precursers but develop directly within the dermis. These are best exemplified by the bones atop our cranium.

The most striking feature of the shark's skeleton is its being made entirely of *cartilage*. It thus exhibits a "fetal" characteristic in remaining cartilaginous during the animal's entire life. Although one may find some areas of ossification due to deposits of calcium salts, they do not form the complex pattern of haversian systems characteristic of the higher vertebrates. The *dermal skeleton* is missing entirely in the shark, except for its placoid scales, teeth and spines.

The best way of studying the shark's skeleton is by the use of mounted specimens of the skeleton. These are usually preserved in fluid or in a solid transparent block of plastic. Mounted skeletons of the entire animal or of its parts such as the skull or the pelvic appendages may be purchased. Although the specimens preserved in fluid may be handled and manipulated, it is best to leave the preparation in the jar, for the material is extremely fragile and delicate and can break easily. Thin parts such as fins may tear from the main skeleton at the slightest touch. Identify the parts of the skeleton preparations by studying the photos and diagrams in this chapter.

The skeleton may be divided into two main areas, the *axial skeleton* and the *appendicular skeleton*.

THE AXIAL SKELETON

In vertebrates, this portion of the skeleton consists of the skull, the vertebral column, and the rib cage.

The Skull—Chondrocranium

Two regions of the skull or chondrocranium are identified, the *neurocranium* and the *splanchnocranium*. Learn to locate and identify the parts listed below.

NEUROCRANIUM — This is the anterior and more dorsal portion of the skull. It protects the brain and the associated sense organs.

Rostrum — The most anterior portion is known as the *rostrum*. It is medially located, tapered anteriorly almost to a point, resembling a bird's beak. It is the support for the shark's snout. Dorsally it is hollowed-out and trough-shaped, forming the *precerbral cavity*, which in life is filled with a gelatinous substance. Ventrally it is keel-shaped, with the keel, the *rostral carina*, extending from about one-half its length posteriorly.

Nasal Capsules — Two large spheres project laterally at the posterior end of the rostrum. These are the *nasal capsules*. Their walls are very thin and therefore often destroyed in the preparation of the skeleton. In the diagrams they are complete and you may see their external openings, the *nares*, ventrally.

Orbits — Further caudally note the large sockets for the eyes, the *orbits*, one on each side, and the *supraorbital crest* dorsally protecting the socket.

Otic Capsules — Behind the eye sockets near the posterior end of the skull, two more large lateral depressions are found on the chondrocranium, the *otic capsules*, one on each side. These contain the inner ears.

Foramina — The chondrocranium is pierced by many perforations, or *foramina*.

> **Epiphysial Foramen** — Mid-dorsally, right behind the rostrum, find the *epiphysial foramen*, through which the *epiphysis*, or *pineal gland*, projects. In many primitive vertebrates the pineal body serves as a third eye.
>
> **Superficial Ophthalmic Foramina** — Also dorsally, above each orbit is a lateral line of about ten foramina, the *superficial ophthalmic foramina*, for the passage of branches of cranial nerves V and VII (the superficial ophthalmic trunk).
>
> **Endolymphatic Fossa** — Continuing mid-dorsally, above the otic capsules, one may find a large depression, the *endolymphatic fossa*. Within this fossa are the anterior paired openings of the *endolymphatic foramina* and the more posterior *perilymphatic foramina*. These pass ventrally, by means of ducts to the inner ears within the otic capsules.
>
> **Rostral Fenestrae** — On the ventral surface the most prominent foramina are the large *rostral fenestrae*, anteriorly located medial to the nasal capsules, on each side of the rostral keel. They open to the cranial cavity.

Notochord — The posterior ventral surface of the skull is flat, covered by a *basal plate*. Along the mid-ventral line, below the surface of the basal plate, one may see a white strand of calcified cartilage. This indicates the position of the *notochord* which persists even in the adult shark. At the anterior end of the notochord, on the ventral surface of the chondrocranium is the *carotid foramen* through which the internal carotid arteries enter the cranial cavity.

Occipital Region — The posterior end of the skull, the *occipital region*, is nearly rectangular in shape. Dorsal to the notochord note the *foramen magnum* which serves as a passageway for the spinal cord out of the cranium. A pair of knuckle-like projections seen ventrally on either side of the foramen magnum are the *occipital condyles*. They serve as articulation points between the chondrocranium and the vertebral column, although fish show very little movement at the occipitovertebral joint. Lateral to the occipital condyles on the posterior edge of the chondrocranium there are two large foramina. The more medial is the *vagus foramen* for the vagus nerve, while the more lateral, the *glossopharyngeal foramen*, serves the nerve of the same name.

Orbit (lateral view) — Looking at the chondrocranium laterally, within the orbital socket, one may see two large foramina, the anterior one the *optic foramen* for the optic nerve and the posterior one, the *trigeminofacial foramen* for the trigeminal and facial nerves. Several smaller foramina for smaller cranial nerves and vessels may also be found. In some preparations one may also see within the orbit, arising from its medial wall a small cartilaginous disc supported by a slender stalk, resembling a golf tee. It is known as the *optic pedicel*. It serves to support the eyeball. A paired process projects at the posterior end of the orbital socket, the *basitrabecular process* which articulates with the upper jaw.

SPLANCHNOCRANIUM — This second portion of the skull is ventral and posterior to the neurocranium. It is called the *splanchnocranium*, or *visceral cranium*, because it consists of the skeletal elements supporting the visceral wall, particularly in the areas of the mouth and the pharynx. The skeletal *visceral arches* form the jaws and the supports for the gill arches.

Visceral Arches — In the jawless fish, *Agnatha*, one finds up to ten visceral arches supporting the mouth parts and gills. They are fused to each other and to the top of the cranium forming a "branchial basket." In the *Squalus acanthias* shark there are only seven.

Mandibular Arch — The first visceral arch, the *mandibular* arch, is modified to form the upper and lower *jaws*. The upper jaw consists of the paired *palatoquadrate* cartilages. It is tied by ligaments to the chondrocranium. The lower jaw consists of the paired *Meckel's* cartilages. They hinge at their caudal ends to form the *angles* of the jaw. A slender pair of *labial cartilages* may be seen on the lateral surfaces of the jaws. They support the lips and help to further strengthen the sides of the jaw. Several rows of sharp, pointed, triangular *teeth*, modified *dermal denticles*, are located at the edges of the jaw. The back rows serve as replacement teeth when the anterior ones are lost or worn out.

Hyoid Arch — The second visceral arch is the *hyoid arch*. It lies immediately posterior to the jaws and extends to the otic capsules. The *spiracles* are located just anterior to the dorsal portion of the hyoid arch. The hyoid arch consists of a single mid-ventral segment, the *basihyal*, which forms the floor of the mouth, and two paired elements, the *ceratohyal* and *hyomandibular* cartilages. The ceratohyal extends dorsolaterally from the basihyal to the angle of the jaw. These elements also serve to support the tongue. Completing the hyoid arch dorsally is the paired hyomandibular cartilage which articualtes with the lateral surface of the otic capsule. The hyoid arch provides primary support for the jaws and is attached to them by ligaments.

The Gill Arches — The last five visceral arches, numbers 3 to 7, posterior to the hyoid arch, are known as the *gill arches*, or branchial arches, because they support the gill elements. They are "U"-shaped, with multi-jointed elements, with the upper (dorsal) portion incomplete. Ventrally, they are all united.

Each gill arch consists of several jointed cartilage elements. The most dorsal is the *pharyngobranchial*, which supports the dorsal portion of the pharynx. They point posteriorly but fail to unite with the vertebrae. The *epibranchial* and *ceratobranchial* segments continue ventrolaterally. Only three short *hypobranchial* elements continue ventrally. These are followed by the most ventral element, the *basibranchial*. There are only two of these in most dogfish, the more caudal one is usually larger and tapers posteriorly. Only the ceratobranchials and the epibranchials bear the *gill lamellae*, the surfaces upon which the exchange of respiratory gases takes place.

Many slender lateral projections are attached to the gill arches (see photos, pages 22 and 23). These are the *gill rays* which support the gill lamellae. Also, one may see shorter, more medial projections, the *gill rakers*. These prevent food from leaving the mouth and pharynx through the gill slits.

Parts of the visceral skeleton persist in the higher vertebrates, even in man. Our six auditory ossicles, parts of the temporal, sphenoid, mandible, and hyoid bones and the cartilage elements of the larynx are vestiges of the visceral skeleton of fishes.

The Vertebral Column and Ribs

In addition to observing the preserved skeletal preparations, you are to do your own dissection of the vertebral bones and ribs. Make a cross sectional cut of the tail between the second dorsal fin and the caudal fin. Your section should appear as the photo on page 25. Also remove a two-inch section of vertebral column just posterior to the first dorsal fin. Remove the skin and muscles to expose the vertebral bones. Refer to the photos and diagrams to help in your identification of parts.

The vertebral column consists of two types of vertebrae: *trunk vertebrae*, those of the main body, and *tail vertebrae*. Let us examine the tail vertebrae first, then see how the more anterior trunk vertebrae differ.

Observe the cross section of the tail and note that the vertebral column lies at the intersection of several *connective tissue septa* which separate the surrounding muscle. These septa will be named in the next chapter on the Muscular System.

TAIL VERTEBRAE — Each vertebra consists of a cylindrical central portion called the *centrum*, which bears a dorsal and a ventral arch of cartilage. The dorsal arch is the *neural arch*. Within this arch, the *spinal cord* passes through the *vertebral canal*. The ventral arch, known as the *hemal arch*, contains the *hemal canal*, through which pass the *caudal artery* and *caudal vein*; the artery dorsal to the vein. The end of each arch tapers to form a spine, the *neural spine* dorsally and the *hemal spine* ventrally.

The Centrum and the Notochord — The notochord runs through the middle of the *centrum*. The part of the centrum immediately adjacent to the notochord has become calcified and appears white. As explained earlier, this calcification, while it strengthens the vertebral column, is not like bone formation in higher vertebrates. Examine two centra in sagittal section. Note that they are shaped like a spool, concave at each end with a gelatinous substance in the core cavity between the two centra. This is also *notochordal tissue* that has persisted from an earlier, embryological stage. This type of centrum is called *amphicoelous*. The sagittal section also reveals that the parts of the centrum near the ends have the broadest notochordal tissue while those at the center have only a narrow band of notochord. This results in the diamond shape of the notochordal tissue within each vertebra.

The Neural Arch — When the *neural arches* of a series of vertebrae are viewed laterally, they seem to be composed of alternating, "V"-shaped, triangular blocks. The block located above the centrum, with its apex pointing dorsally, is known as the *neural plate*. The other block, located in the joint between the centra, wedged between the neural plate elements, with its apex pointing ventrally, forms the *dorsal intercalary plate*. Each plate is perforated by a foramen. The dorsal roots of spinal nerves exit through foramina in the dorsal intercalary plate, while the ventral roots exit through the foramina in the neural plates. Find these in your specimen. Much smaller, *ventral intercalary plates* are found ventrally between vertebrae.

The Hemal Arch — The *hemal arch*, below the centrum, is composed of repeated blocks of cartilage, the *hemal plates*. In the lateral wall of the hemal canal are *foramina* through which branches of the caudal artery and vein exit.

TRUNK VERTEBRAE — The structure of the trunk vertebrae is basically the same as the tail vertebrae, lacking, however, the hemal arches. Instead they have short, ventrolateral processes, *basapophyses*, that project from the sides of the septum of each vertebra. Small *rib* cartilages, extending horizontally, are attached to these projections.

THE APPENDICULAR SKELETON

The *appendicular skeleton* refers to the cartilages of the *pectoral* and *pelvic* girdles and to their respective *fins*.

Pectoral Girdle and Pectoral Fins

Study the preserved skeletal preparations and the accompanying photographs.

PECTORAL GIRDLE — A "U"-shaped structure which encircles the ventral side of the trunk. It consists of a median ventral *coracoid bar* which articulates laterally with the *scapular* and *suprascapular cartilages*. These have a depression, the *glenoid fossa*, for articulation with the *pectoral fin*.

PECTORAL FINS — They consist of three *basal fin cartilages* which articulate with the girdle at the glenoid surface. From anterior to posterior they are the *propterygium, mesopterygium,* and *metapterygium*. These articulate distally with many slender *radial cartilages*. The most distal portions of the pectoral fins are broad, yet thin and flexible to allow for free movement. These are the *ceratotrichia*. The supporting cartilages of the basal and radial components are collectively known as the *pterygiophores* (supporters of the fin).

Pelvic Girdle and Pelvic Fins

PELVIC GIRDLE — The *pelvic girdle* of the dogfish shark is composed of a single transverse rod of cartilage, the *puboischiac bar,* located in the ventral abdominal wall just anterior to the cloaca. At each end of the bar there extends a short dorsolateral process, the *iliac process*.

PELVIC FINS — The *basal cartilages* of the *pelvic fins* consist of only two elements, the *propterygium* and the much longer *metapterygium* which extends caudally. As in the pectoral fins, there are a series of *radial cartilages* which articulate with the basal cartilages. Again, the most distal portions of the fins are known as *ceratotrichia*. In male sharks a highly modified radial cartilage forms the *clasper,* by which sperm are transferred to the cloaca of the female.

Dorsal and Caudal Fins

DORSAL FINS — The *dorsal fins,* one anterior and the other more posterior, although not a part of the appendicular skeleton, are here described because they are similar in structure to the pectoral and pelvic fins. The larger *basal cartilages* are next to the vertebral column, the smaller *radial cartilages* are more distal. Finally, the thin parallel most distal rays are the *ceratotrichia*. A sharp pointed *spine* projects dorsally from the anterior end of each dorsal fin.

CAUDAL FIN — One may also find *ceratotrichia* on the caudal fin. These articulate directly with the neural and hemal arches. The end of the vertebral column turns upward into the dorsal part of the caudal fin. This type of tail is termed *heterocercal*.

THE SKELETON – VENTRAL VIEW

- Chondrocranium
  - Rostrum
  - Nasal Capsule
- Jaws (Mandibular Arch)
  - Palatoquadrate Cartilage
  - Meckel's Cartilage
- Visceral Arches
  - Gill Rays
- Pectoral Girdle and Fins
  - Coracoid Bar
  - Pectoral Fin
- Vertebral Column
- Spine
- Dorsal Fin (Posterior)
- Puboischiac Bar

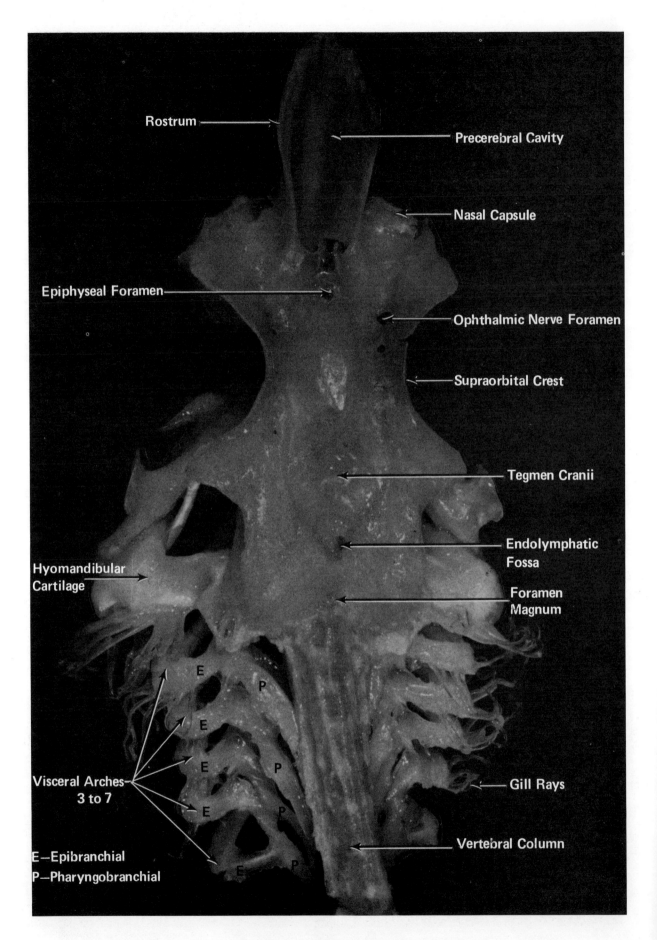

SKULL – DORSAL VIEW

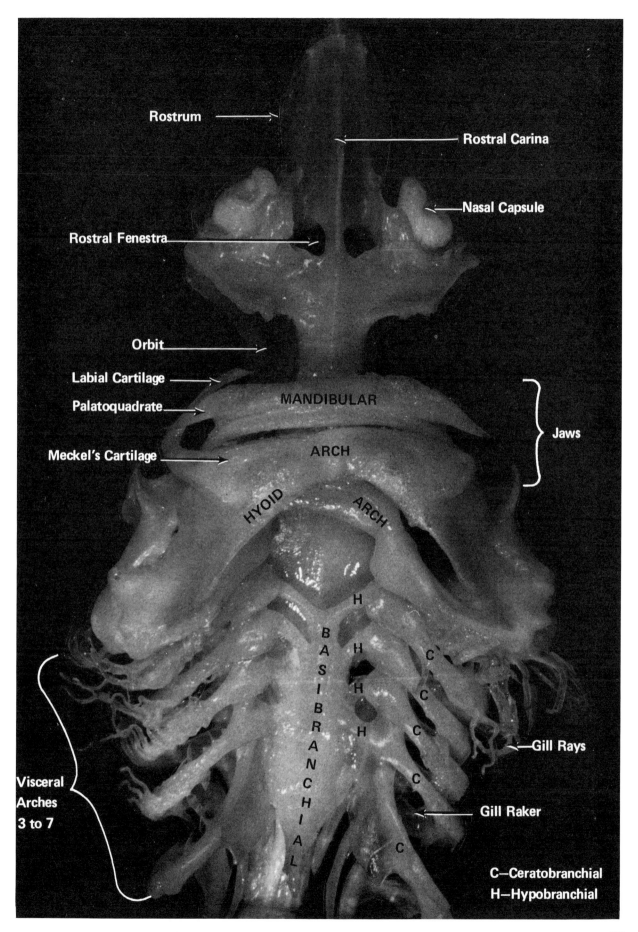

SKULL – VENTRAL VIEW

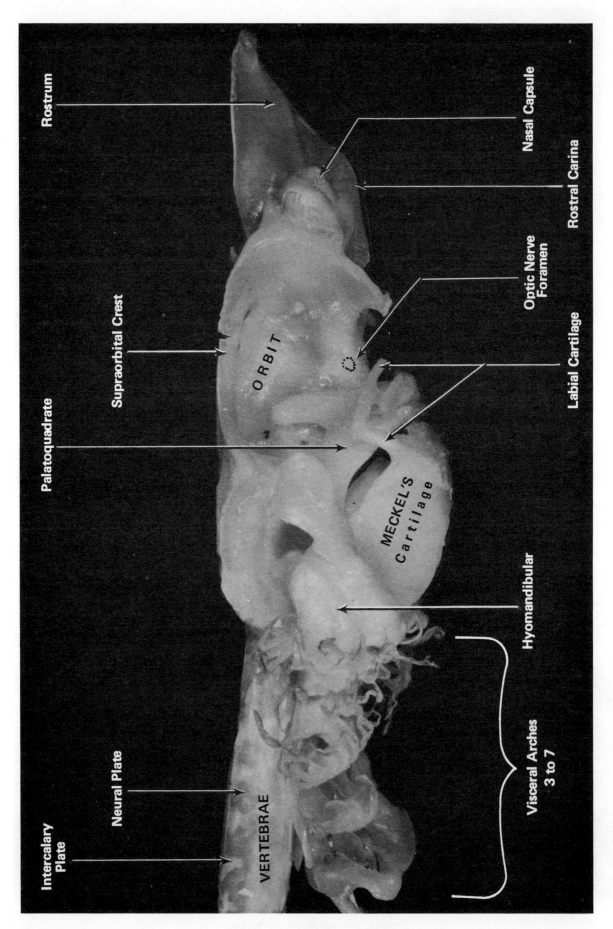

SKULL – LATERAL VIEW

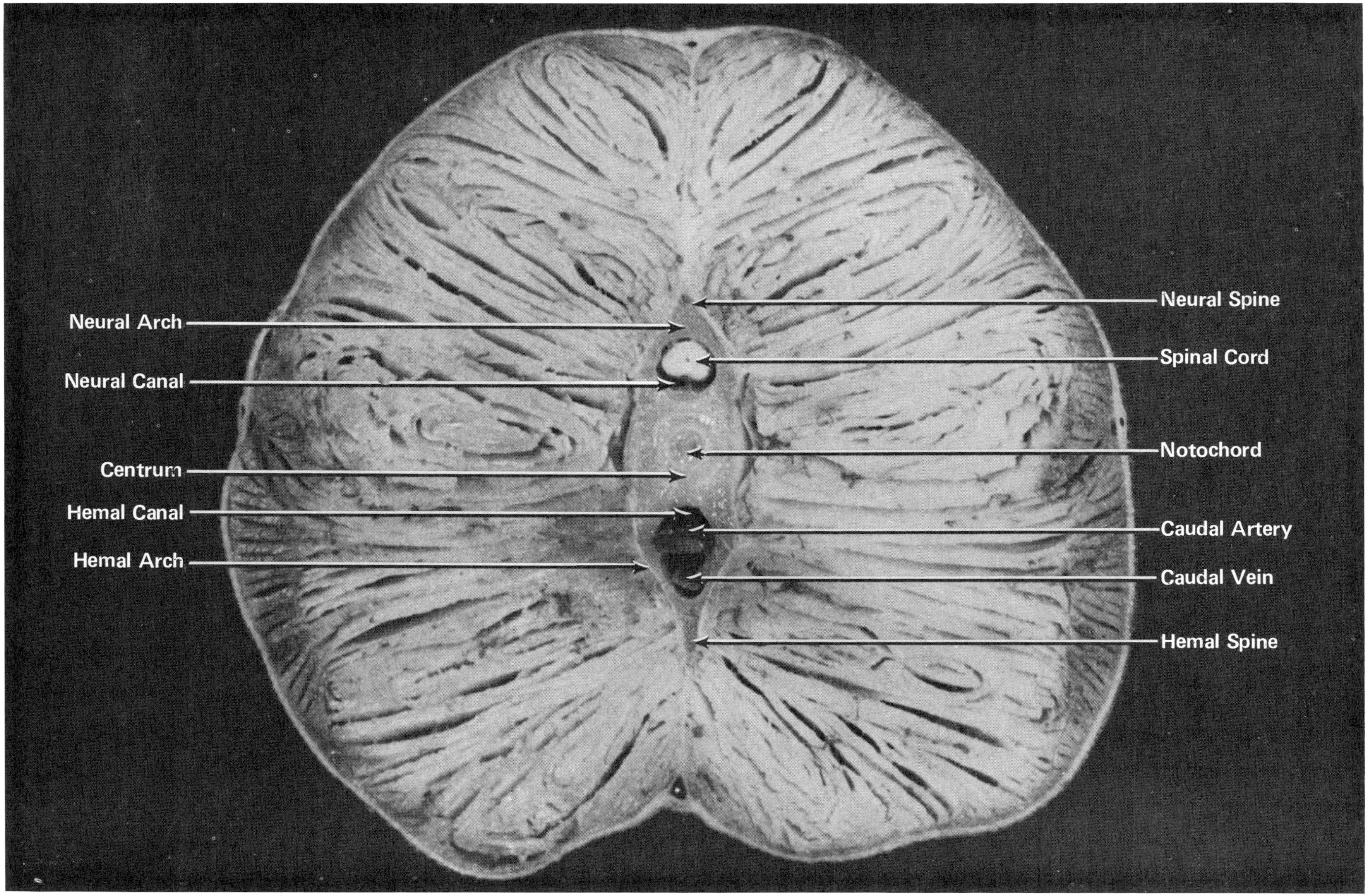

Neural Spine
Spinal Cord
Notochord
Caudal Artery
Caudal Vein
Hemal Spine

Neural Arch
Neural Canal
Centrum
Hemal Canal
Hemal Arch

CAUDAL VERTEBRA – CROSS SECTION

25

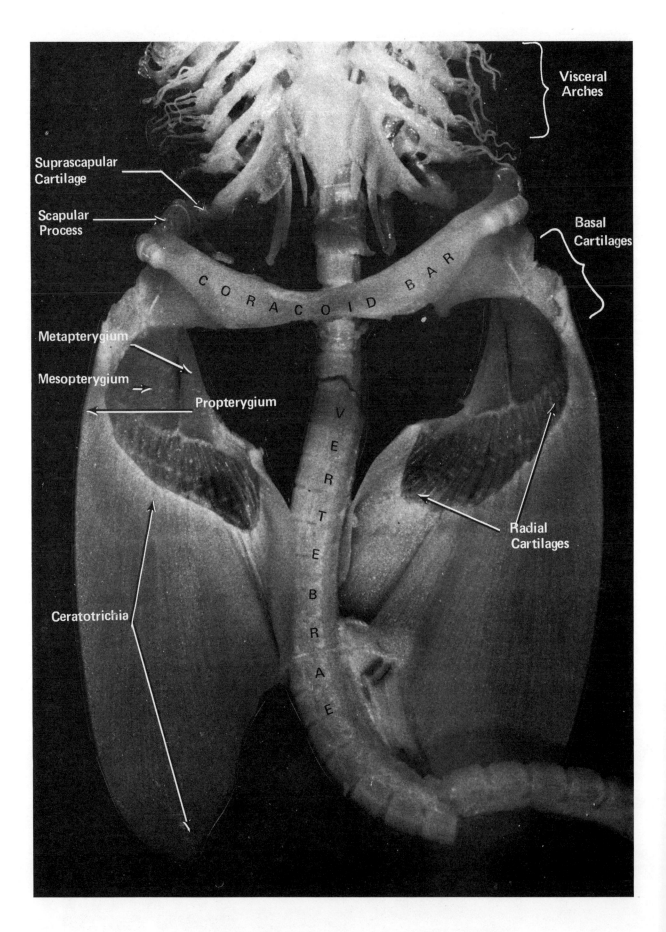

PECTORAL GIRDLE AND FINS — VENTRAL VIEW

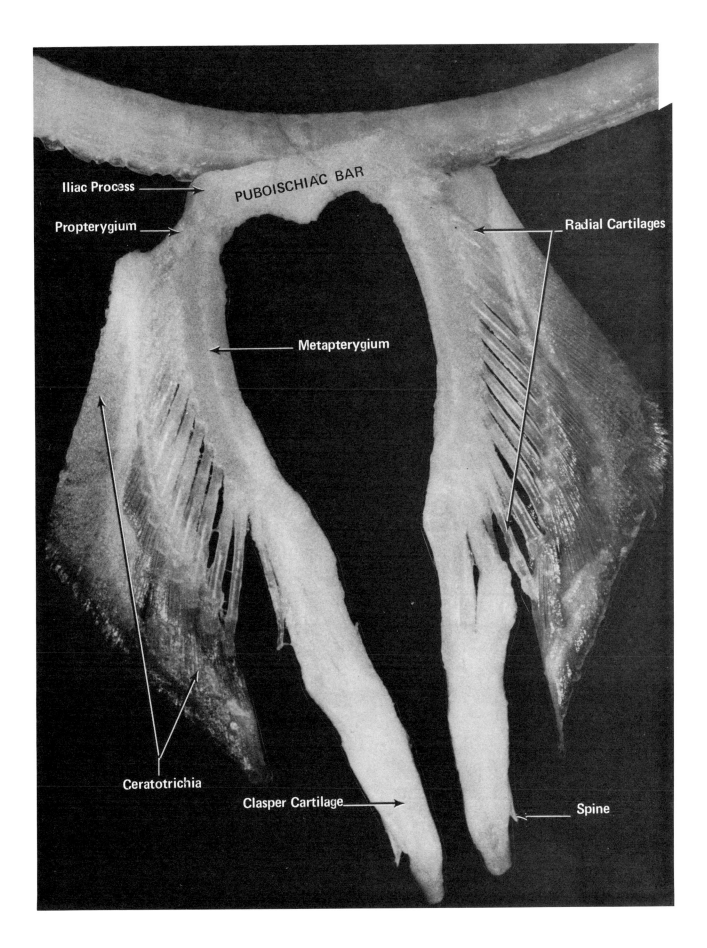

PELVIC GIRDLE AND FINS, MALE — VENTRAL VIEW

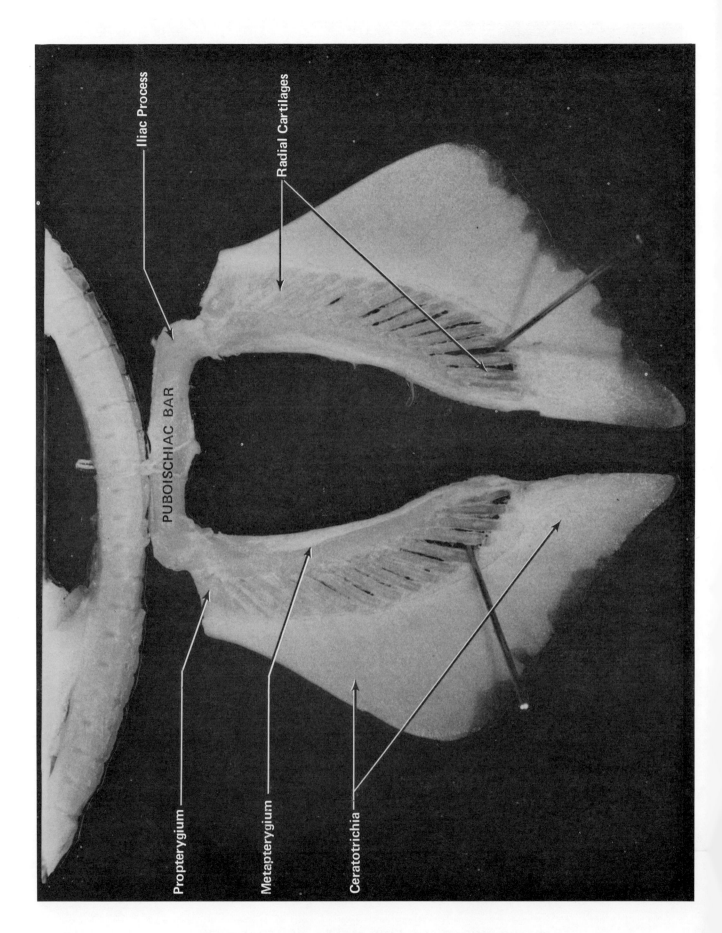

PELVIC GIRDLE AND FINS, FEMALE – VENTRAL VIEW

Name _____ Section _____ Date _____

SELF-QUIZ I
THE SKELETAL SYSTEM

1. Differentiate between the terms chondrocranium and splanchnocranium.
2. Name the skeletal components of the gill arches in a dorsal to ventral direction.
3. Name five fossa and/or foramina of the chondrocranium.
4. Name the skeletal components of the mandibular and hyoid arches.
5. What is the relationship between the notochord and the vertebral column?
6. Name the skeletal components of the pectoral and pelvic fins.
7. Name and locate all of the fins of the shark.
8. Name the structures seen in a cross-sectional view of a caudal vertebra.
9. What advantages does a cartilaginous skeleton have over a bony skeleton?
10. Define each of the terms listed below.

ANSWERS

1. _____
2. _____
3. _____
4. _____
5. _____
6. _____
7. _____
8. _____
9. _____
10. a. rostrum _____
    b. Meckel's cartilage _____
    c. otic capsule _____
    d. foramen magnum _____
    e. coracoid bar _____
    f. basibranchial cartilage _____
    g. metapterygium _____
    h. intercalary plate _____
    i. hemal arch _____
    j. radial cartilages _____

Label all of the features indicated on the following illustration.

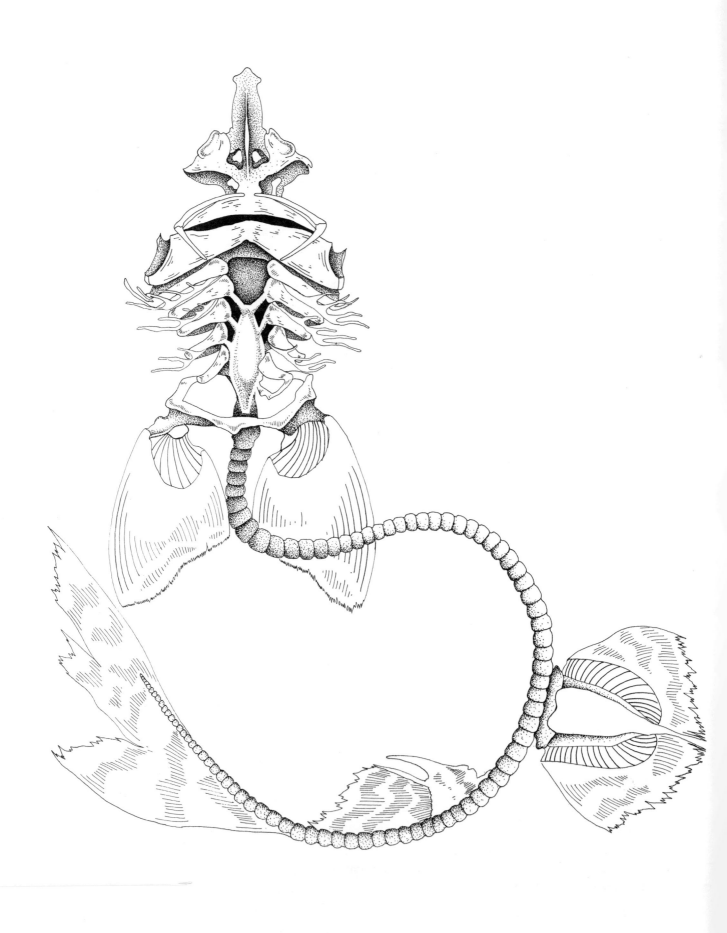

THE SKELETON — VENTRAL VIEW

THE MUSCULAR SYSTEM

INTRODUCTION

Many of the introductory remarks concerning the muscular system, while concerned primarily with the dogfish shark, are equally true for the higher vertebrates.

The muscles of *Squalus* are a good example of the musculature in primitive vertebrates. In the higher forms these have been modified by migrations, splitting, fusion, or a combination of factors. In the shark the natural groups of muscles can be recognized, identified and studied in the adult, while in higher forms the original natural muscle groups can only be found in the embryo.

SKELETAL MUSCLES enable the body to move. They are involved in moving the entire shark through the water as well as in moving individual visceral structures such as the jaws and gill arches.

Most muscles are firmly anchored to the skeleton at one end, the *origin* of the muscle, while the other end is attached to the skeletal element to be moved, and is known as the *insertion*. The fleshy central portion is termed the *belly*. The ends of a muscle are attached to the skeleton most often by means of a narrow band of connective tissue called a *tendon*. They may also be joined directly to the *periosteum* of a bone. Finally muscles may be united with each other or with a skeletal element by means of a broad, flat sheet of tendinous tissue known as an *aponeurosis*.

As you dissect, locate the origins and insertions of the muscles studied. Then free the muscle from other muscles and from the nerves and blood vessels associated with it. The fine, transparent connective tissue which binds adjacent muscles is *deep fascia*, while tougher and more fibrous *superficial fascia* connects the skin to the muscles below. When the muscle has been freed, pull it gently. This will duplicate the muscle's normal contraction. Observe which bones or organs are moved and which remain relatively stable.

The *action* of a muscle results from its contraction. Muscles are usually arranged in *antagonistic* pairs. This means that while a muscle will cause a structure to move in one direction, its antagonist will cause it to move in the opposite direction.

ACTIONS OF MUSCLES

Flexion — to bend at a joint decreasing the angle at that joint; examples: elbow or knee joint.

Extension — to straighten joint increasing the anle at that joint.

Adduction – to move appendage toward sagittal midline; example: lowering arms from shoulder level to rest at sides.

Abduction – to move appendage away from sagittal midline; example: raising arms from rest at sides to shoulder level.

Supination — to turn palm of hand upward.

Pronation — to turn palm of hand downward.

Rotation — to move a structure about a point; example: turning head from side to side.

Circumduction — when the distal end of a limb describes a circle while the proximal end remains fixed, as the vertex of a cone; example: the movement of the extended arm in drawing a circle on the blackboard.

The study of the shark musculature will be divided into three aspects. We shall first study the major body muscles which propel the fish through the water, then those of the gill (branchial) area and the head, and finally, those of the fins (appendicular muscles).

BODY MUSCULATURE

The Dissection

Remove a section of skin in order to observe typical body musculature.

Proceed as follows.

Make a very shallow incision into the skin at the mid-dorsal line, directly posterior to the second dorsal fin. Continue to cut caudally for about two inches. At each of the two ends, cut the skin ventrally along the sides of the body till you reach the mid-ventral line. Do not cut too deeply for you may destroy the muscles you wish to study. Use a blunt instrument such as a probe, the handle of your scalpel, even your fingers, to remove the section of skin whose perimeter you have just cut. If the shark's skin adheres very tightly to the underlying musculature, the use of a scalpel may be necessary. Then, use your scalpel to make a clean cross-sectional cut, through the entire body of the shark, cutting off the tail, directly posterior to the second dorsal fin. This affords a view of the transverse as well as the lateral arrangement of muscle bundles.

Identify the parts in your dissection as described below. Consult the accompanying photos. See photo, page 42.

MYOTOMES (MYOMERES) — The muscles you have exposed are composed of segments termed *myotomes*. They are arranged in a zigzag, "W"-shaped pattern along the entire length of the animal's trunk and tail. Only in the gill region are they partially interrupted. The myotomes are separated from one another by connective tissue partitions called *myosepta*. The dorsal portion is clearly separated from the ventral portion by the *horizontal (transverse) septum,* a band of connective tissue between the muscle bundles. This septum extends lateraly toward the position of the *lateral line* on the outer surface of the trunk. The bundles of myotomes, dorsal to the horizontal septum, are known as *epaxial,* while those ventral are known as *hypaxial*. Along the mid-ventral region, from coracoid bar to anus, a band of white connective tissue separates the myotomes of the right and left sides. This is the *linea alba.* The direction of the fibers of each myotome is longitudinal on either side of the horizontal septum and somewhat oblique near the more dorsal and ventral extremes.

Examine the cross-sectional view of the muscle bundles and identify two more connective tissue septa. The *dorsal median septum* lies between the right and left epaxial bundles. The *ventral medial septum* between the right and left hypaxial bundles in the region caudal to the anus. Recall that anterior to the anus, the linea alba separates right and left hypaxial bundles.

MUSCLES OF THE BRANCHIAL AND HEAD REGIONS

The epaxial and hypaxial muscle bundles of the trunk and tail are interrupted by specialized muscles in the area of the gills (branchial), and below the pharynx (hypobranchial). They are involved with opening and closing the gill pouches during the exchange of respiratory gases and in opening and closing the jaws.

The Dissection

Remove the skin of the dogfish as follows: Start mid-ventrally at the pectoral fins and proceed anteriorly to the lower jaw. Continue to cut and remove the skin laterally toward the right side. Remove all of the skin covering the gills, below and posterior to the eye around the spiracle then dorsally to the top of the head. Take care to avoid removing the superficial muscles.

After the skin has been removed, observe the regions of the head and gills in lateral view.

The muscles in this region are involved with feeding and respiration, opening and closing the jaws, swallowing, and passing water across the gills by compressing and expanding the gill pouches. They are arranged in series, corresponding to the visceral arches which they serve.

These *branchial muscles* are divided into three groups:

1. **Superficial constrictors**
2. **Levators**
3. **Interarcuals**

A fourth group, the **hypobranchials,** form the floor of the pharynx and will be viewed from the ventral surface. They help in opening the mouth, swallowing, and expanding the gill pouches.

Find and identify the muscles described on your specimen.

SUPERFICIAL CONSTRICTORS — As their name indicates, they are primarily involved in constricting the gill muscles and compressing the gill pouches. Their action is, however, modified in visceral arches 1 and 2 where they move the jaws.

First Dorsal Constrictor — This is composed of three muscles:

1. **Adductor Mandibulae (Quadratomandibularis)**
2. **Spiracularis (Craniomaxillaris)**
3. **Preorbitalis (Suborbital)**

   Adductor Mandibulae (Quadratomandibularis) — This easily seen large mass of muscle is located at the angle of the jaws. It arises from the posterior portion of the palatoquadrate cartilage and inserts on Meckel's cartilage.

   Spiracularis (Craniomaxillaris) — This small muscle lies on the anterior wall of the spiracular valve. It lies posterior to one of the larger levator muscles (to be described later). It has been questioned whether it is a separate muscle or a part of the larger levator.

   Preorbitalis (Suborbital) — This is a deeper muscle, cylindrically shaped, which lies between the upper jaw and the eye. It originates in the mid-ventral surface of the cranium and inserts on Meckel's cartilage.

First Ventral Constrictor

   Intermandibularis — This is a broad, thin, ventral constrictor muscle of the first visceral arch. It lies posterior to the mouth. It originates at the mid-ventral raphe of Meckel's cartilage and inserts at its ventral borders. It acts to raise the floor of the mouth and thus to force water out of the gill slits.

Second Constrictor — This is composed of two muscles:

1. **Hyoid Constrictor**
2. **Interhyoid**

Hyoid Constrictor — This muscle lies in the region of the second, or hyoid, arch and is wider than the others. It extends from the first gill slit to the angle of the jaw. It includes both dorsal and ventral fibers.

Interhyoid — This ventral muscle is the second of the hyoid arch constrictors. It is the anterio-ventral extension of the hyoid constrictor. It is broad and thin and lies near the ventral midline. In order to expose this muscle, it will be necessary to dissect a part of an overlaying muscle, the intermandibularis, a constrictor of the first visceral arch.

Third to Sixth Constrictors — Beginning with the area just dorsal to the gill slits, find the *dorsal constrictors,* and ventral to the gill slits the *ventral constrictors*. They are numbered 3 to 6 and correspond to the gill arches of the same number. These constrictors are similar to each other. A white vertical connective tissue band, the *raphe,* extends above and below each gill slit and separates adjacent constrictor sets. They overlap each other anteriorly; thus, each is partly hidden by the constrictor anterior to it.

LEVATORS — The *levator* branchial group consists of three distinct muscles. As the name indicates, they raise the jaws and the five gill arches.

First Levator — *Levator Palatoquadrati (Levator Maxillae Superioris)* — This muscle is located anterior to the *spiracle*. It arises from the side of the otic capsule and inserts on the palatoquadrate cartilage.

Second Levator — *Levator Hyomandibulae (Hyoid Levator)* — The second levator muscle is located caudal to the first, in back of the spiracle. It also originates on the otic capsule and inserts on the lateral surface of the hyomandibular cartilage. It also acts to raise the jaws.

Third to Sixth Levators — *Cucullaris* — This muscle acts as unified levator of all five gill arches. It lies dorsal to the superficial gill constrictors and is triangular in shape. It originates from the epibranchial musculature in the occipital region of the cranium to insert posteriorly on the pectoral girdle and the last gill arch. Another function of this muscle is to move the pectoral girdle and fin cranially and dorsally. For this reason it is also considered to be partly homologous to the *trapezius* muscle of higher vertebrates.

INTERARCUALS — These are a series of small muscles which act upon the gill arch cartilages. To see them it is necessary to cut dorsally into the gill pouches, separate the epaxial and cucullaris muscles and to expose the pharyngobranchial cartilages.

Dorsal (Medial) Interarcuals — These muscles extend between adjacent pharyngobranchial cartilages. They pull the gill arches cranially.

Lateral Interarcuals — These muscles extend from the pharyngobranchials to the adjacent epibranchial cartilages.

Subspinal — This muscle originates on the posterior portion of the cranium near the foramen magnum and inserts on the first pharyngobranchial cartilage. It, together with the dorsal interarcuals, draws the gill arches cranially.

Branchial Adductors — These small muscles extend between the epibranchial and ceratobranchial cartilages. They act to flex the branchial arches.

HYPOBRANCHIAL — Turn the shark ventral side up. These muscles are locate between the coracoid bar of the pectoral girdle and Meckel's cartilage (lower jaw). To see them all it is necessary to remove the superficial intermandibular, interhyoid muscles as well as the superficial ventral constrictors. They act in opening the mouth, swallowing, and expanding the gill cartilages.

Common Coracoarcuals — This muscle pair lies immediately anterior to the coracoid bar, from which it originates. Its fibers taper anteriorly and are continuous with those of the *coracohyoids, coracomandibular,* and *coracobranchials*.

Coracohyoids — This muscle pair originating from the *common coracoarcuals* continues cranially on either side of the mid-ventral line. It inserts on the basihyal, the ventral median cartilage of the hyoid arch.

Coracomandibular — A narrow muscle lying in the mid-ventral line between the two *coracohyoids*. It also originates from the *common coracoarcuals* and inserts upon Meckel's cartilage.

Coracobranchials — These muscles are deeper than the other hypobranchials. Five segments pass to the cartilages of the gill arches. They act to expand the pharyngeal cavity. The most posterior slip forms a part of the anterolateral wall of the pericardial cavity.

EPIBRANCHIAL — In the dorsal part of the shark the epaxial muscles, uninterrupted by the gills, continue their body pattern of segmentation till the chondrocranium.

APPENDICULAR MUSCLES

In fish the pattern of appendicular muscles is very simple. The fins do not undergo complex movements. The primary forward thrust is achieved by the movements of the body and the tail. The fins are for steering and maintaining stability.

Fin Musculature

PECTORAL FIN — Remove the skin of the pectoral fins; from both its ventral and dorsal surfaces. Also remove some of the skin immediately anterior to the fin.

Flexor and Extensor (Adductor and Abductor, Depressor and Levator) — You will find a single ventral and a single dorsal mass of muscle radiating toward the distal end of the fin. These are the *pectoral flexor* on the ventral surface and the *pectoral extensor* on the dorsal surface. The ventral flexor depresses the fin and pulls it forward, while the dorsal extensor raises the fin and pulls it posteriorly.

The ventral pectoral flexor originates on the coracoid bar and inserts on the pterygiophores of the fin, while the dorsal pectoral extensor originates on the scapular process of pectoral girdle and also inserts on the pterygiophores.

PELVIC FIN — The muscles of the pelvic fin are somewhat more complex than those of the pectoral fin. In addition, in males, part of the pelvic fin is modified as a *clasper* for the transfer of sperm to the female.

Remove the skin of one of the pelvic fins of a female shark from both its ventral and dorsal surfaces. Also remove some of the skin immediately anterior to the fin.

Flexor and Extensor (Adductor and Abductor, Depressor and Levator) — The muscle mass on the ventral surface of the fin, the *flexor*, may be divided into the *proximal pelvic flexor* muscle, and the *distal pelvic flexor* muscle. The proximal portion arises from the puboischial bar and inserts on the metapterygium; the distal portion originates from the metapterygium and inserts on the radial cartilages.

The dorsal muscle mass, the *extensor*, arises from two origins. The more *superficial extensor* originates from the iliac process of the puboischiae bar, while the *deeper extensor* originates from the metapterygium. Both insert upon the radials and the ceratotrichia.

The muscles of the pelvic fins of males are fundamentally the same as in females. However, some portions of the dorsal and ventral muscle mass extend into the male's clasper as separate muscles. Also, in the male pelvic fin the ventral flexor cannot be fully exposed until a long muscular sac, the *siphon*, is reflected.

DORSAL FINS — Although these fins are not ordinarily considered appendicular or movable, they possess *radial muscles* upon their sides.

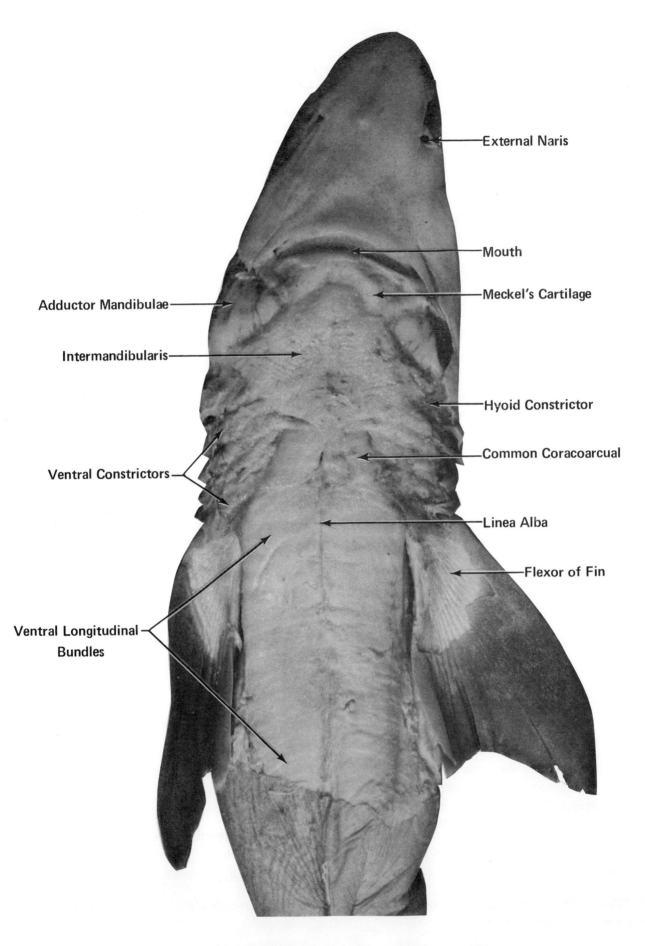

ANTERIOR MUSCULATURE – VENTRAL VIEW

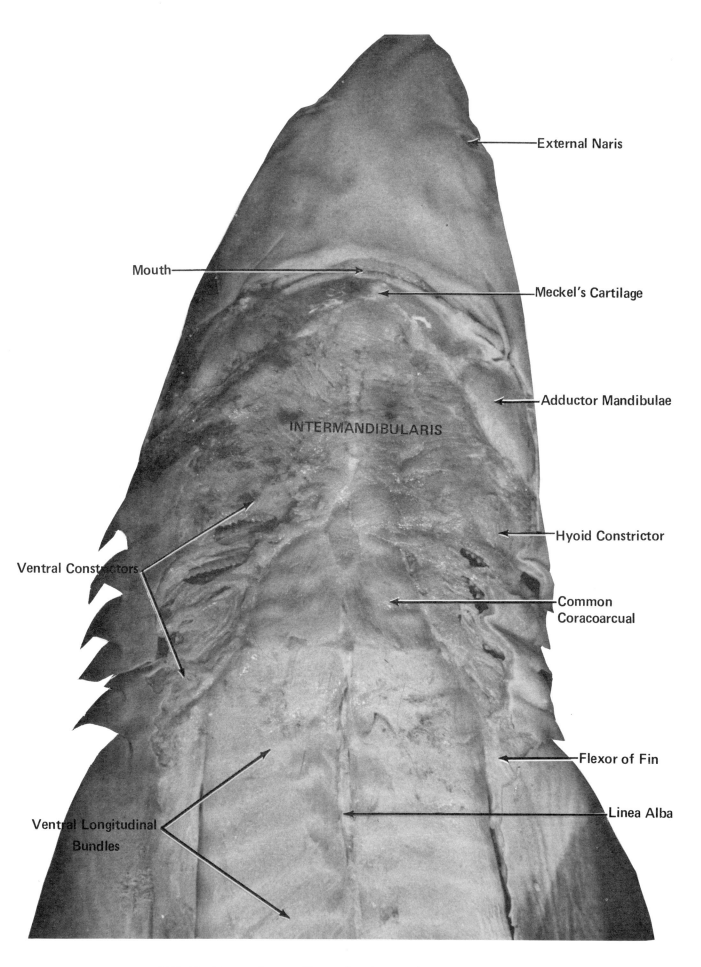

ANTERIOR MUSCULATURE (CLOSE-UP) — VENTRAL VIEW

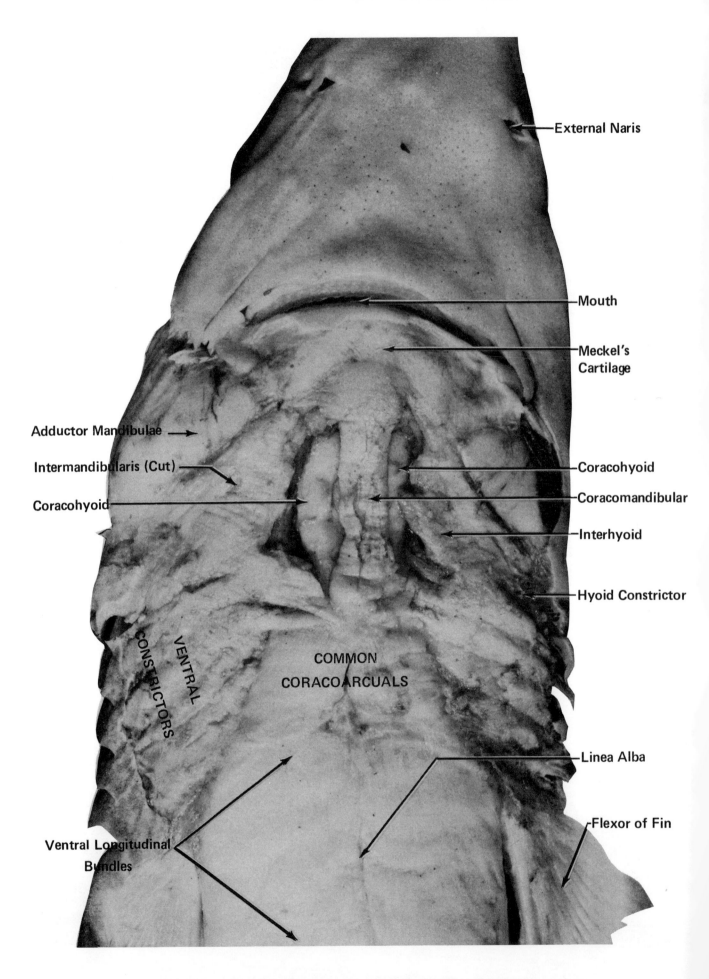

38 ANTERIOR MUSCULATURE, DEEPER MUSCLES — VENTRAL VIEW

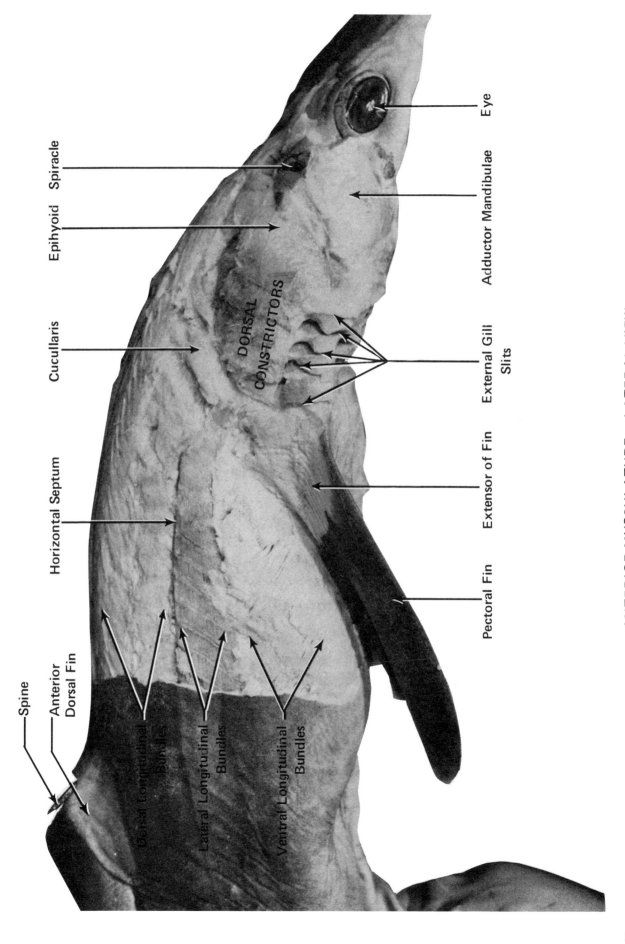

ANTERIOR MUSCULATURE – LATERAL VIEW

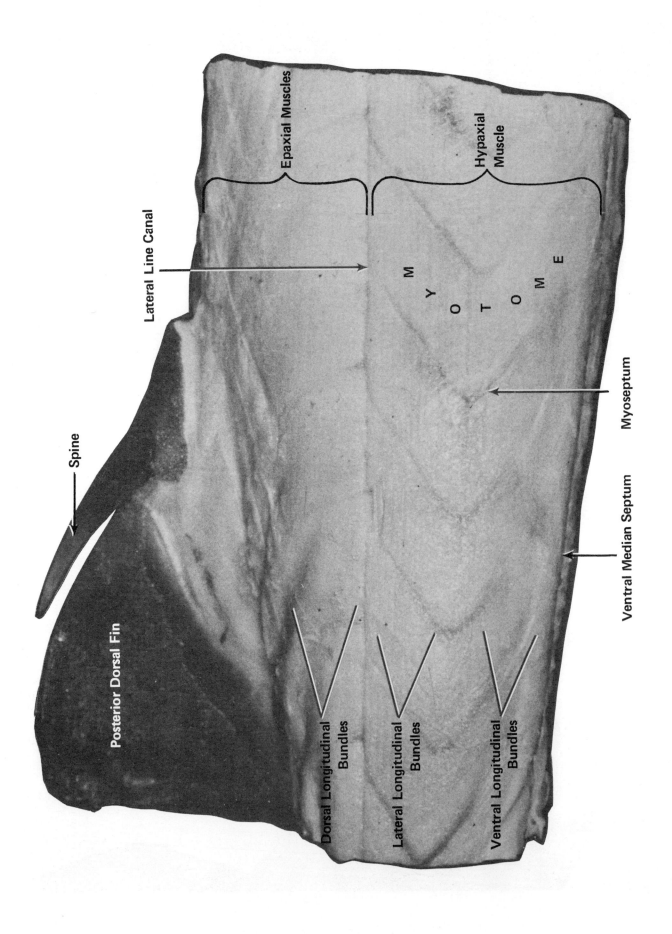

POSTERIOR MUSCULATURE – LATERAL VIEW

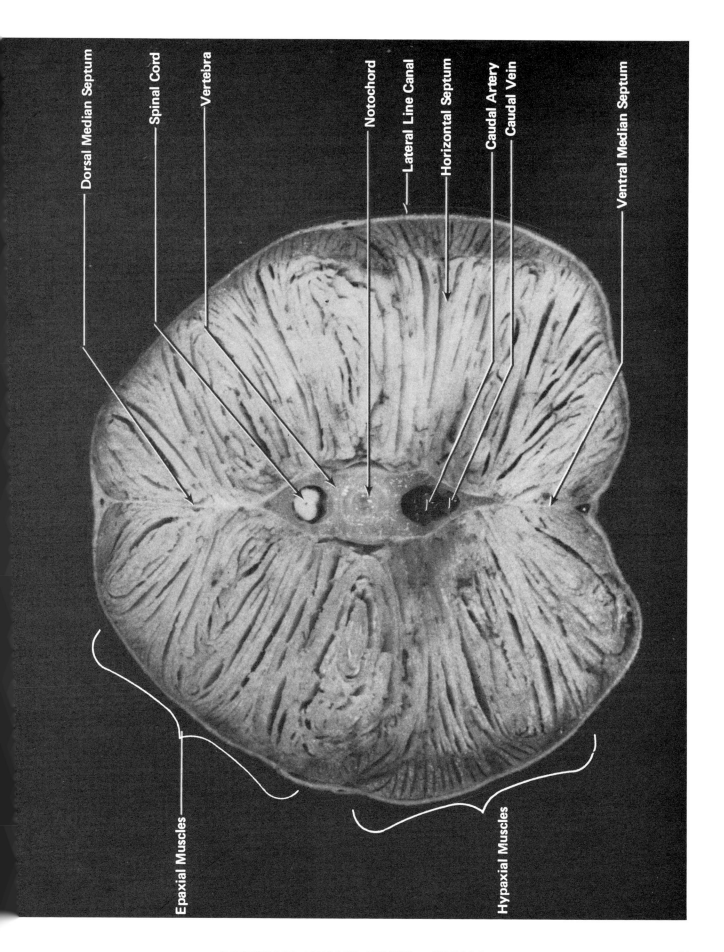

POSTERIOR MUSCULATURE – CROSS SECTION

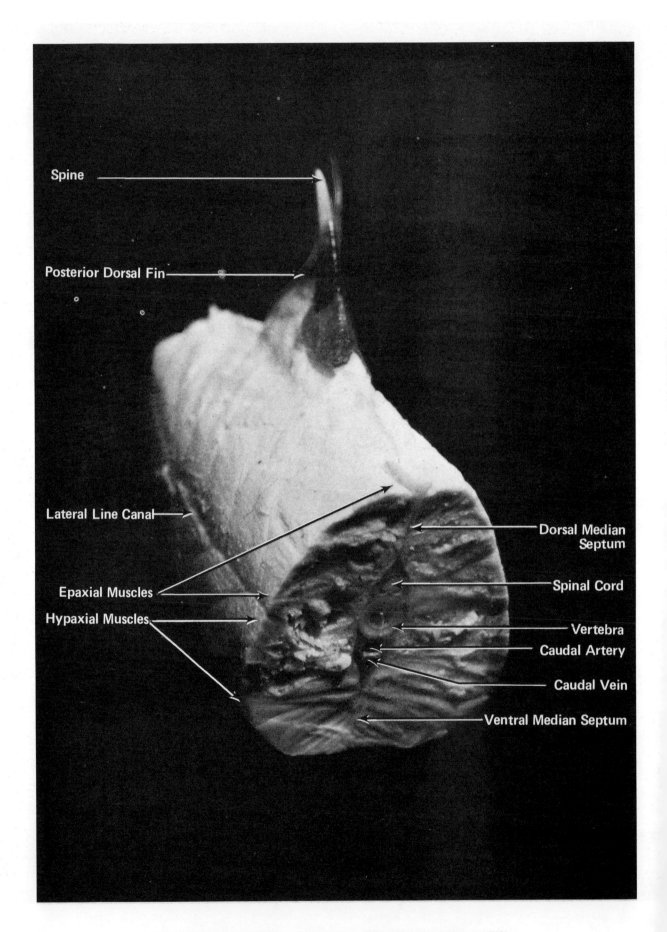

POSTERIOR MUSCULATURE — STEREOSCOPIC VIEW

Name _____ Section _____ Date _____

SELF-QUIZ II
THE MUSCULAR SYSTEM

1. Differentiate between epaxial and hypaxial muscles.
2. Name some of the muscles that move the jaw.
3. Name some of the muscles that move the gill arches.
4. Name four hypobranchial muscles.
5. The skeletal unit to be moved by a muscle serves as the (a) origin, (b) rotation, (c) insertion, (d) action, for that muscle.
6. How do the body musculature and tail musculature differ in cross-section?
7. Which muscles move the fins?
8. Which are the major muscles that help propel the shark through the water?
9. Name the muscles that move the eye.
10. Define each of the terms listed below.

ANSWERS

1. _____
2. _____
3. _____
4. _____
5. _____
6. _____
7. _____
8. _____
9. _____
10. a. myotome _____

    b. levator muscle _____

    c. transect _____

    d. interarcual muscle _____

    e. linea alba _____

    f. depressor _____

    g. antagonist _____

    h. raphe _____

    i. constrictor _____

    j. adductor _____

Label all of the features indicated on the following illustration.

43

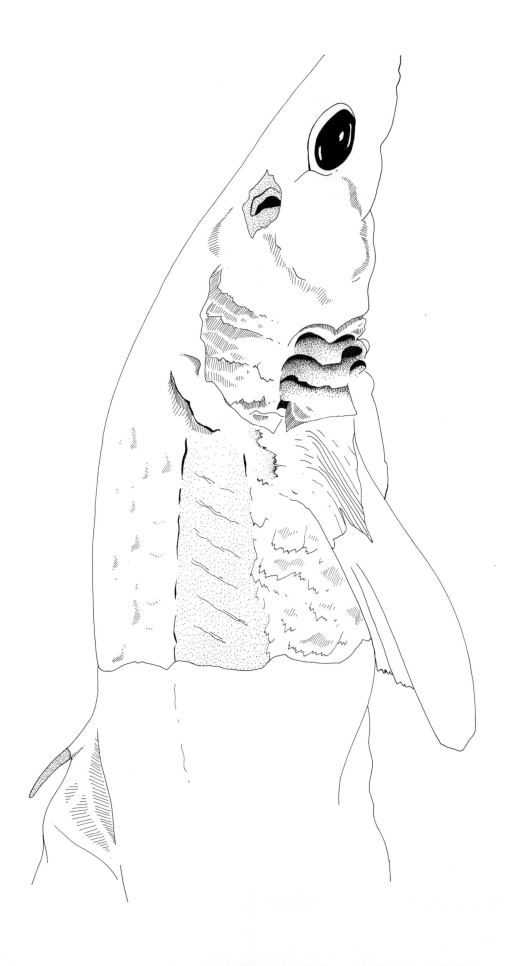

ANTERIOR MUSCULATURE – LATERAL VIEW

THE DIGESTIVE AND RESPIRATORY SYSTEMS

In the shark these two systems are studied as one since the mouth and pharynx serve both as organs of digestion and respiration. We shall begin our dissection with the organs that are primarily related to the process of digestion.

The Dissection

Turn your specimen ventral side up. Make a mid-ventral incision just anterior to the *cloacal opening*. Cut through the skin and muscle in an anterior direction slightly to the right of the mid-ventral line. Continue your cut to the *coracoid bar* of the pectoral girdle. At that point use your scissors and proceed with the blunt end to cut the skin and muscles laterally toward the right and to the left. Similarly, at the point you began the dissection, near the cloacal opening, cut laterally to the right and to the left. You have thus exposed the large body cavity known as the *pleuroperitoneal cavity*. Fold back the large flaps of body wall you have cut and secure them with large dissection pins.

PLEUROPERITONEAL CAVITY

COELOM — The *coelom* or body cavity of the shark is divided into the larger posterior chamber, the *pleuroperitoneal cavity*, and the smaller anterior *pericardial cavity* which contains the heart. The two cavities are separated by a partition, the *transverse septum*.

Recognize and identify on your specimen all of the structures listed. Make use of the labeled photographs.

PERITONEUM — A smooth, shiny membrane will be seen lining the inside of the body wall. This membrane is the *parietal peritoneum*. The membrane covering the surface of the visceral organs is the *visceral peritoneum*. As you move some of the visceral organs to the side, you will see that they are suspeneded dorsally by a double membrane of peritoneum know as *mesentery*. Different sections of mesentery have various names indicating the types of organ suspended. These will be named as the organs are discussed.

LIVER — The largest organ lying within the pleuroperitoneal cavity is the liver. Its two main lobes, the *right and left lobes,* extend from the pectoral girdle posteriorly most of the length of the pleuroperitoneal cavity. A third lobe, the *median lobe,* is much shorter than the others, and as the name indicates, is located medially. Locate the elongated sac, the green *gall bladder* along the right edge of the median lobe. The *common bile duct* extends from the anterior portion of the gall bladder to the duodenum.

The anterior portion of the liver is attached to the ventral body wall by a membrane, the *falciform ligament*, and to the transverse septum by the *coronary ligament*.

The great bulk of the liver can be visualized when compared to other organs. A giant 20-foot basking shark which weighed a total of 13,850 pounds had a 1,850-pound liver. The liver is rich in oil. This is the form in which the shark stores energy, not as fats. The oil's specific gravity is also responsible for giving the shark a limited amount of buoyancy, although it cannot keep him afloat as does the swim bladder of bony fish.

ESOPHAGUS — Move the large lobes of the liver laterally to reveal other organs of the body cavity. You will see a thick muscular tube extending from the top of the cavity at the mid-line posteriorly toward the left. This is the *esophagus*. It passes through the transverse septum to connect the oral cavity and pharynx with the stomach.

STOMACH — The esophagus leads into the "J"-shaped *stomach*. The upper portion, the cardiac region, continues as the *main body*, and ends at the duodenal end. The left-hand outer border of the stomach is called the *greater curvature* while the right-hand, inner border is the *lesser curvature*. Dorsally the stomach is supported by a membrane, a derivative of the mesentery, the *mesogaster (greater omentum)*. Another membrane, the *lesser omentum (gastrohepatoduodenal ligament)*, supports the stomach ventrally.

Cut the stomach open along its long axis. Avoid the large blood vessels seen externally. Note its contents. It will generally consist of partially digested remains of fish, squid, or other sea animals. Wash out the inside of the stomach under slowly running water. Note the *mucosa*, the inner lining membrane. The longitudinal folds, the *rugae*, help in the churning and mixing the food with digestive juices. A circular muscular valve, the *pyloric sphincter*, is located at the distel end of the stomach. It regulates the passage of partially digested food out of the stomach.

DUODENUM — A short "U"-shaped tube, the *duodenum*, the first portion of the small intestine, connects the stomach to the next part of the alimentary canal. The bile duct from the gall bladder enters the dorsal surface of the duodenum.

PANCREAS — Ventral to the duodenum and partially obscuring it is the whitish glandular tissue of the *pancreas*. The greater portion of the pancreas is not seen until one examines the dorsal surface of the stomach and duodenum. Here the dorsal elongated segment of the pancreas may be found. Connecting the dorsal and ventral lobes of the pancreas is the *isthmus*, a slender band of pancreatic tissue. The secretions of the pancreas enter the duodenum by way of the *pancreatic duct*.

SPLEEN — Near the posterior end of the stomach find the dark, triangular-shaped *spleen*. Although not a part of the digestive system but the lymphatic system, it is closely associated with the digestive organs of vertebrates. A part of the mesogaster membrane extends between the spleen and the stomach, the *gastrosplenic ligament*, which ties these two organs together.

VALVULAR INTESTINE — This second, and much larger, portion of the small intestine follows the duodenum. Its outer surface is marked by rings. This hints at the contour to be found within. Cut away the outer tissue of this portion of the alimentary canal. Exercise caution in not injuring the blood vessels which are located on the surface of the intestine. Wash out the contents. You will see a symmetrical spiral shape within, the *spiral valve*. It adds surface area for digestion and absorption to an otherwise relatively short intestine. In higher vertebrates, increases in surface area are accomplished by means of coiling and projecting finger-like villi.

COLON — This narrowed continuation of the valvular intestine is located at the posterior end of the pleuroperitoneal cavity. If the end of the colon has been everted through the cloacal opening, pull it back into the body cavity.

RECTAL GLAND (DIGITIFORM GLAND) — A slender, narrowed, finger-like structure, the *rectal gland*, closed at one end, leads into the colon by means of a duct. It has been shown to excrete salt (NaCl) in concentrations higher than that of the shark's body fluids or sea water. It is thus an organ of *osmoregulation*, regulating the shark's salt balance. The *mesorectum*, a section of the dorsal mesentery, attaches the colon and rectal gland to the mid-dorsal line.

CLOACA — This last portion of the alimentary canal collects the products of the *colon* as well as the *urogenital* ducts. This catch-all basin leading to the outside by means of the *cloacal opening* has rightly deserved its name which means sewer. In higher vertebrates, separate exits exist for the rectum (anus), for the urinary bladder (urethra), and for the reproductive system (vagina).

ABDOMINAL PORES — The coelomic cavity of higher vertebrates is closed and has no direct connection with the outside. In the shark, however, a pair of *abdominal pores* may be found posterolateral to the cloacal opening. Pass a blunt probe through the pores to confirm the connection between the coelom and the outside. In some specimens the lips of the pore may have grown together. Their function has not yet been determined.

NON-DIGESTIVE ORGANS — Several organs, not part of the digestive system, may be seen in the pleuroperitoneal cavity. Most are part of the reproductive and genital systems.

The *gonads (testes* or *ovaries)* may be found by moving the liver and digestive organs to one side. They are located in the anterodorsal portion of the body cavity. The supporting mesentery of the testes is the *mesorchium,* of the ovaries, the *mesovarium*. An additional mesentery, the *mesotubarium,* supports the oviducts of mature females.

The *kidneys* are dark elongated structures, running the length of the body cavity on either side of the mid-dorsal line. In mature animals the *oviducts* in females, and the *archinephric* ducts in males, may be seen running from the gonad to the cloaca.

All of these urogenital structures will be studied in more detail in a later chapter.

ORAL CAVITY AND PHARYNX

Although the mouth, the oral cavity, and the pharynx of the shark serve as passageways for food, they play a more active role in respiration.

RESPIRATION — Water taken into the mouth and pharynx passes over the gill filaments, through the gill slits, to the outside. During this process, oxygen is removed and transported into the circulatory system and carbon dioxide is released from the blood at the gill lamellae and exits via the gill slits.

The Dissection

With the shark lying ventral side up, insert the blunt blade of a strong pair of scissors into the right corner of the shark's mouth. Begin cutting posteriorly through the angle of the jaws across the gill slits as far back as the pectoral girdle. Cut across the ventral musculature to lay the entire preparation flat on your dissection pan. Your dissection should appear as in the photo on page 54. Secure the specimen with large dissection pins. Locate and identify each of the following on your specimen.

The Oral Cavity

This is the area enclosed by the jaws (mendibular arch) and the hyoid arch.

TEETH — These triangular sharp structures are arranged in several rows beginning at the outer edges of the upper and lower jaws. They are similar to the dermal denticles found on the skin of the shark in their structure and development. Behind the visible rows of teeth are other rows within the mucosa, usually folded downward ready to replace any lost. It has been estimated that the mouth of the great white shark may contain 400 teeth!

TONGUE — The *tongue* of the shark is different from the true tongue of higher vertebrates. It is practically immovable, without muscles under the epithelium. It is supported anteriorly by the basihyal cartilage of the hyoid arch and posteriorly by the pharyngeal arch cartilages. These can be palpated by the fingertips.

Pharynx

The *pharynx* is the portion of the alimentary canal posterior to the hyoid arch between the gill slits. Posteriorly it narrows to form the *esophagus*.

SPIRACLES — The *spiracles* are openings in the anterior roof of the pharynx, in its dorsolateral wall. The shark can bring water into its pharynx to the gills by way of the spiracle even when its mouth is closed. Pass a blunt probe into one spiracle and note where it exits.

GILLS — The *gills* are the respiratory organs of the shark. They are composed of *gill lamellae, blood vessels,* and *supporting cartilaginous structures*.

As you look at the pharynx you will see five internal *gill slits*. They lead into cavities called *gill pouches*, which lead to the outside by *external gill slits*. The gill slits are supported by cartilaginous *gill arches* and guarded by small cartilaginous papillae-like *gill rakers* which act as strainers to prevent food particles from leaving the pharynx through the gill slits.

Branchial Bars — The partitions between gill pouches are referred as branchial bars. They include all of the following gill elements: the *interbranchial septa, anterior and posterior demibranchs, gill arch cartilages, gill rakers, gill rays, branchial blood vessels,* the *adductor muscle* and the *valve flap*.

Gill Lamellae and **Interbranchial Septa** — Look closely at the cut surfaces of the gills. Note the *primary gill lamellae (gill filaments)*. They are pink colored in injected specimens. They are a radially folded, highly vascularized tissue. They are attached to the surface of a tough connective tissue, the *interbranchial septum*.

Examine the primary gill lamellae with a hand lens and note that each bears many small, closely packed *secondary lamellae* extending perpendicularly from the surface of the primary lamellae.

Each septum is attached medially to a portion of the cartilaginous gill arch. Distally they extend beyond the gill lamellae, to end upon the superficial constrictor muscles which act as flap-like *valves* to open and close the external gill slits.

Demibranchs and Holobranchs — The gill lamellae on one side of a branchial bar are called a *demibranch (hemibranch)*, or half gill. The demibranchs on the anterior and posterior surface of a single branchial bar are termed a *holobranch*, or complete gill. Thus, one holobranch belongs to two different gill pouches; the anterior half (demibranch) to the anterior gill pouch, the posterior gill demibranch to the posterior gill pouch. Note: There is no demibranch (gill filament) on the posterior wall of the last (fifth) gill pouch.

Pseudobranchs — Examine one of the spiracles. Use your scissors to cut into the spiracle. Examine its walls and the flap-like spiracular valve. A minute demibranch called a *pseudobranch*, or false gill, may be found on the anterior wall of the spiracle. Since oxygenated blood passes through the pseudobranch, it does not have any respiratory function. It is conjectured that it may have other functions such as excretory or endocrine.

Gill Rays — These are fine cartilagionous, finger-like, lateral extensions of the gill arch cartilages. They may be found by scraping off the gill lamellae from the surface of the supporting septa. The *gill rays* provide support for the interbranchial septa.

Branchial Blood Vessels — The gills are provided with a rich blood supply. Near the proximal end of each septum locate a single *afferent branchial artery*, which brings deoxygenated blood to the gill lamellae, in the middle of the septum, and two *efferent branchial arteries*, on either side of the afferent vessel, which carry oxygenated blood from the gills. A single *adductor muscle* lies medial to each gill arch. It compresses the gill cartilages.

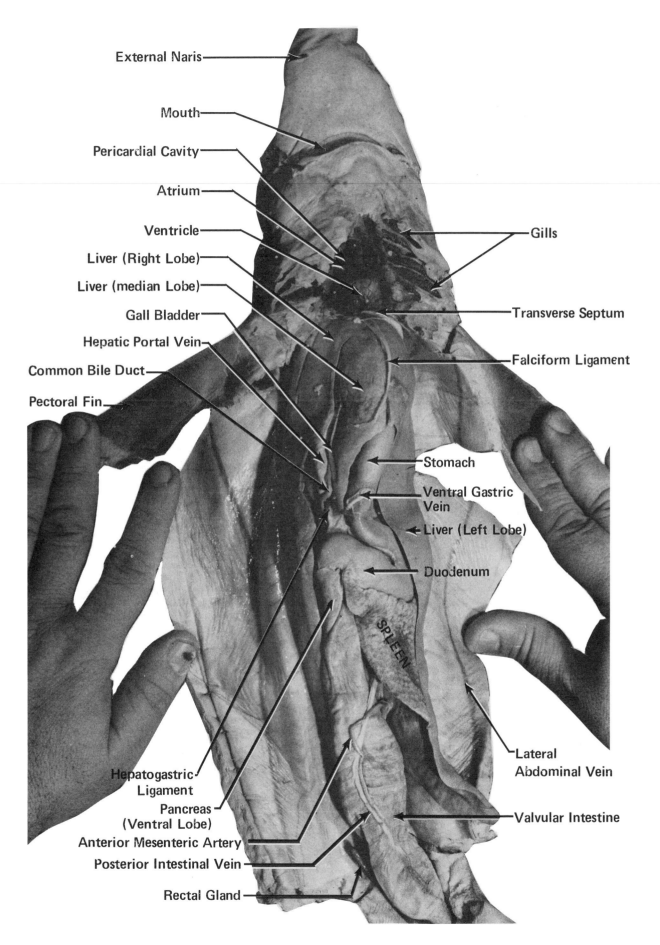

THE VISCERA – VENTRAL VIEW

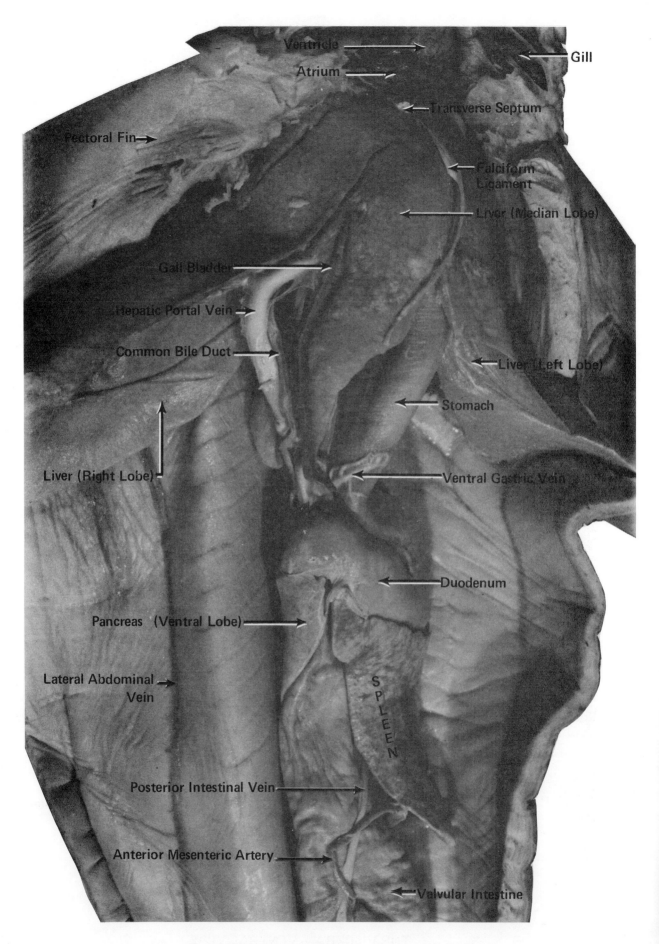

THE VISCERA (CLOSE-UP) — VENTRAL VIEW

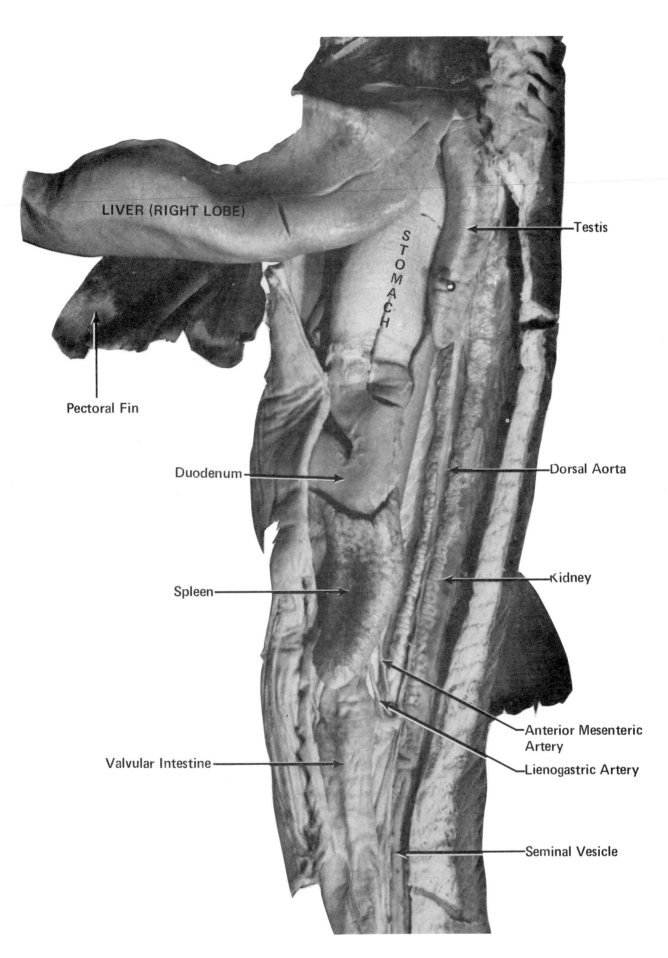

THE DORSAL VISCERA

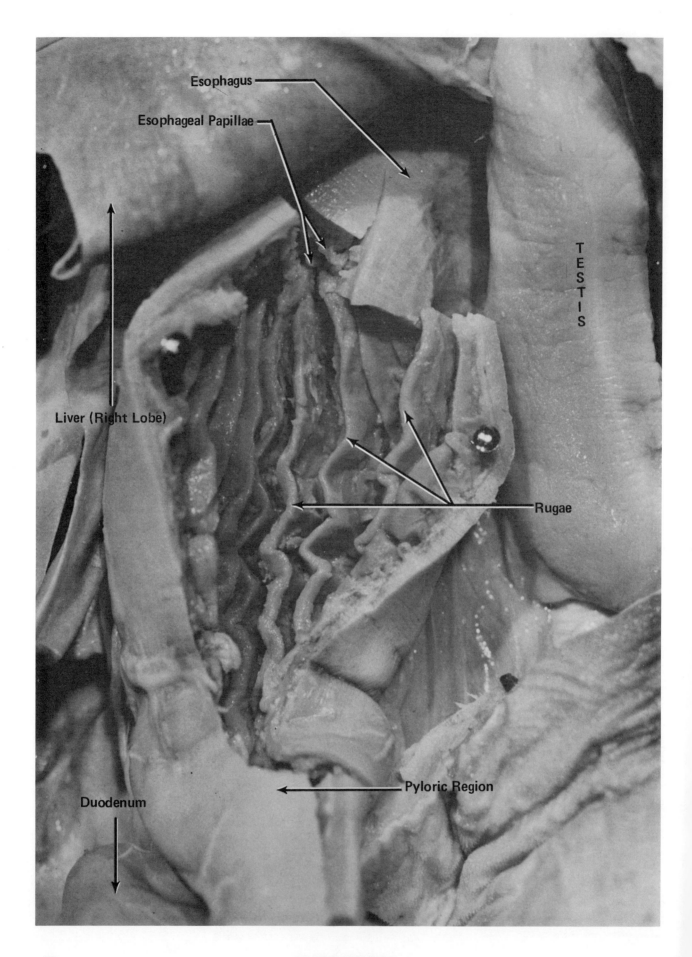

52 THE STOMACH

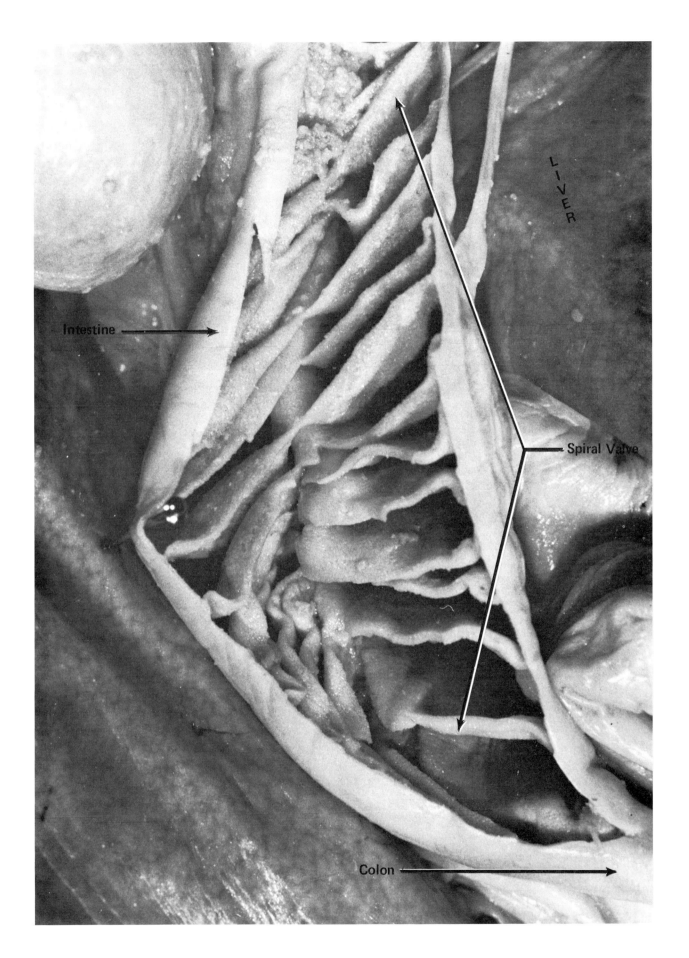

THE VALVULAR INTESTINE

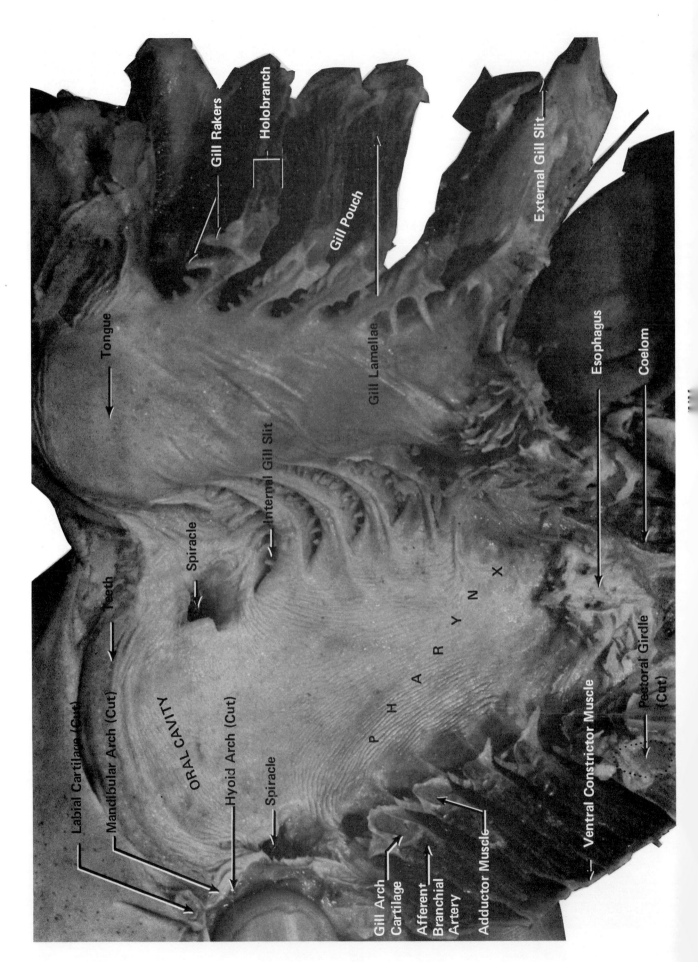

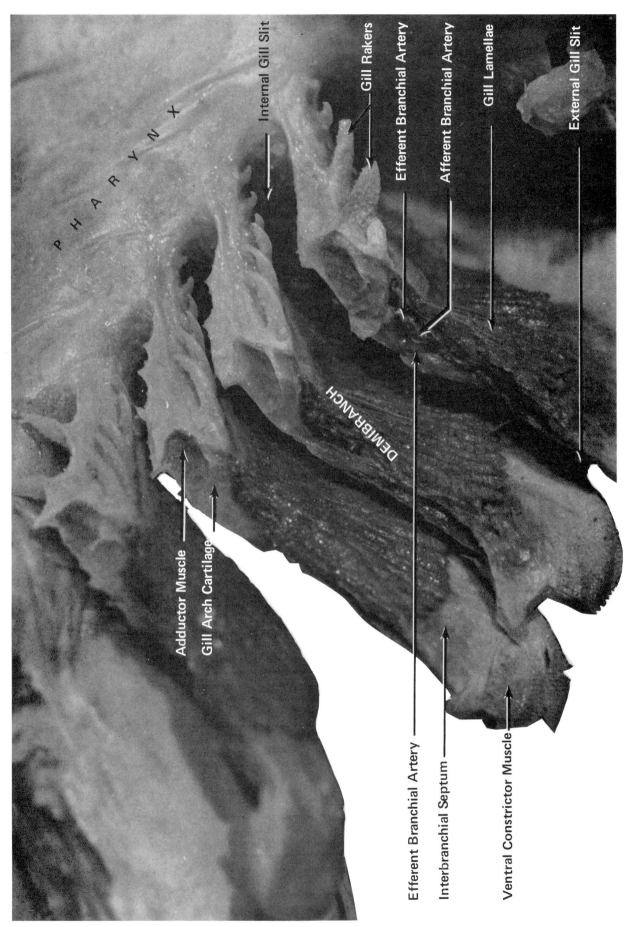

THE GILLS (CLOSE-UP)

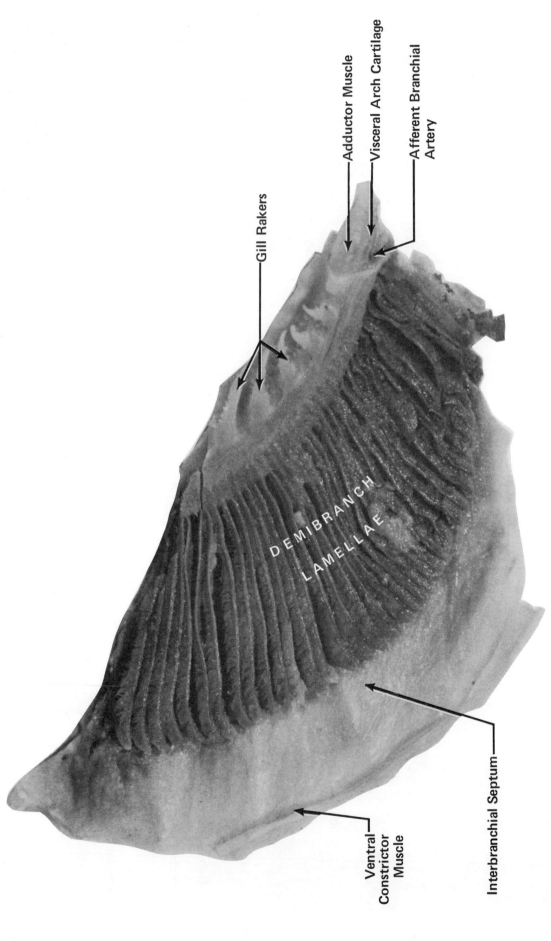

GILL SURFACE

Name _____ Section _____ Date _____

SELF-QUIZ III
THE DIGESTIVE AND RESPIRATORY SYSTEMS

1. Name the lobes of the liver.
2. Name the various openings of the pharynx.
3. Describe the parts of a single branchial bar.
4. What is the function of the spiracle in respiration?
5. Name three sections of the dorsal mesentery.
6. What is the advantage of the spiral valve in the intestine?
7. What is the function of the rectal gland?
8. How do the gills function in respiration?
9. Name the sections of the stomach.
10. Define each of the terms listed below.

ANSWERS

1. _____
2. _____
3. _____
4. _____
5. _____
6. _____
7. _____
8. _____
9. _____
10. a. pyloric valve _____
    b. rugae _____
    c. gill pouch _____
    d. demibranch _____
    e. holobranch _____
    f. falciform ligament _____
    g. esophageal papillae _____
    h. internal gill slit _____
    i. interbranchial septum _____
    j. gill raker _____

Label all of the features indicated on the following illustration.

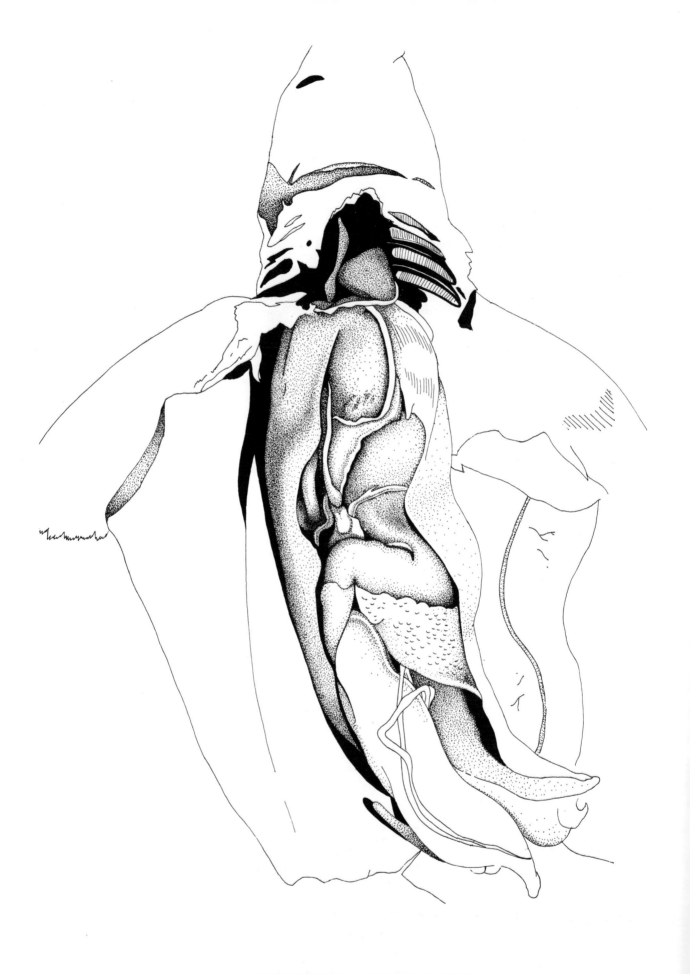

THE VISCERA — VENTRAL VIEW

THE CIRCULATORY SYSTEM

The circulatory system is involved in transporting substances to and from the body cells. It consists of the heart, the arteries, veins, sinuses, capillaries, and the blood.

This extensive system will be studied in several phases. The first is the heart.

THE PERICARDIAL CAVITY

The *pericardial cavity* is the upper portion of the *coelom*, the body cavity. It is much smaller than the lower coelom, the pleuroperitoneal cavity, which we studied earlier. It is located anterior to the transverse septum and contains the heart and the major blood vessels leading to and from the heart.

The Dissection

Place the shark ventral surface upward. Locate the *pectoral girdle*. If not done in a previous dissection, remove the skin anterior to the coracoid bar, till the edge of the lower jaw (Meckel's cartilage). Remove the ventral hypobranchial musculature in this area. A membrane will be found covering a triangular cavity, the *pericardial cavity*. Remove the membrane to expose the heart and some of its major blood vessels. Locate and identify all of the parts listed below.

PERICARDIUM — This is the membrane lining the inner walls of the pericardial cavity. It is known as the *parietal pericardium*. The layer of membrane covering the heart is the *visceral pericardium*. It is fused with the heart and cannot be peeled off. At the upper and lower borders of the heart, observe where the parietal and visceral pericardia join and are continuous with one another.

THE HEART

The shark heart is composed of four distinct continuous tube-like chambers. Blood is passed from the more posterior end anteriorly in sequence, from one chamber to the next. The four chambers are:

1. **sinus venosus**
2. **atrium**
3. **ventricle**
4. **conus anteriosus**

SINUS VENOSUS — This is the most posterior of the four chambers. Deoxynaged blood from the entire body returns first to this chamber of the heart. Lift the main portion of the heart and observe a broad, thin-walled, flattened, almost horizontal, sac-like structure extending the width of the pericardial cavity. Its base lies upon the transverse septum.

ATRIUM — This chamber is anterior and dorsal to the sinus venosus. It is also thin-walled with two lateral bulging lobes. It receives blood from the sinus venosus.

VENTRICLE — This most ventral part of the heart is first seen upon exposing the pericardial cavity. It is an oval-shaped, thick-walled, muscular sac, lying ventral to the atrium. Paired *coronary arteries* may be seen on its ventral surface as well as on the conus arteriosus.

CONUS ARTERIOSUS — A thick, muscular, tubular structure which originates from the anterior surface of the ventricle. It extends anteriorly to the upper end of the paricardial cavity.

Note: Unlike the heart of higher vertebrates, the heart of the shark transports *deoxygenated blood* only. The process of oxygenation takes place at the gills, from where blood passes to the entire body without first returning to the heart.

The Dissection

As you open the heart chambers you will find coagulated blood and, in injected specimens, you will also find rubber-like colored latex. Remove these materials with blunt probes and wash out the heart chambers under gently running water.

SINUS VENOSUS — Probe the inner walls of this chamber and locate the lateral openings for the *common cardinal veins* bringing blood from the entire body and posteriorly the *hepatic sinuses* from the liver which enter medially. Other veins enter anteriolaterally bringing blood from the head.

SINOATRIAL APERTURE — Find this opening between the sinus venosus and the atrium. It is regulated by a pair of *sinoatrial valves*.

ATRIOVENTRICULAR APERTURE — This opening between the atrium and the ventricle is guarded by a pair of *atrioventricular valves*. They prevent blood from returning from the ventricle to the atrium.

VENTRICLE — Note the thick muscular walls of the ventricle and the comparatively small inner ventricular cavity. The inner muscular surface is irregularly folded.

CONUS ARTERIOSUS — Along the walls of the conus arteriosis count a series of *semilunar valves*. The two proximal sets of valves are smaller than the more anterior distal one.

THE VENTRAL AORTA AND AFFERENT BRANCHIAL ARTERIES

The Dissection

Continue the dissection from the pericardial cavity anteriorly. Remove the ventral *hypobranchial muscles* and connective tissues until you reach the lower jaw, Meckel's cartilage. Trace the *conus arteriosus* anteriorly. Follow the major branching blood vessels.

THE VENTRAL AORTA — After the conus arteriosus exits the anterior end of the pericardial cavity, it continues as the *ventral aorta*. Trace it anteriorly. Be careful since it is generally not injected. It gives off five pairs of *afferent branchial arteries* which carry *deoxygenated blood* from the heart to the gills.

AFFERENT BRANCHIAL ARTERIES — These arteries pass laterally from the medial ventral aorta carrying deoxygenated blood to the gills. The *fourth and fifth afferent branchial arteries* arise together from a common branch of the ventral aorta just anterior to the pericardial cavity. Shortly beyond that the ventral aorta gives off the *third afferent branchial* artery. At the level of the hyoid arch the ventral aorta bifurcates, passes laterally, gives off two branches on each side, the *first and second afferent branchial arteries*. Trace these afferent vessels to the gill arches. Note that the second, third, fourth, and fifth afferent branchial arteries enter the interbranchial bars and serve the holobranchs of similarly numbered gill arches. The first afferent interbranchial artery supplies only a demibranch on the anterior surface of the first gill pouch.

THE EFFERENT BRANCHIAL ARTERIES

The *efferent branchial arteries* serve to return oxygenated blood from the gills. This blood is then distributed to all parts of the body.

The Dissection

You have already examined and studied the oral cavity and the pharynx (see page 47 and photo page 54). If you have not already done so, with the shark lying ventral side up, insert the blunt blade of a strong pair of scissors into the right corner of the shark's mouth. Begin cutting posteriorly through the angle of the jaws, across the gill slits as far back as the pectoral girdle. Cut across the ventral musculature to lay the entire preparation flat on your dissection pan. Secure the specimen with large dissection pins.

Remove the mucous membrane from the roof of the mouth and pharynx. Find and identify the large arteries and their branches in the area of the gill arches.

Also remove the mucous membrane covering the floor of the mouth and pharynx. This will expose the dorsal aspect of the heart and the afferent branchial arteries already studied. Your dissection should now appear as the one in the photo on page 68.

EFFERENT BRANCHIAL ARTERIES — Four pairs of arteries may be seen arising from the gills and uniting in the midline to form the median *dorsal aorta*. Careful dissection will reveal the source of each efferent branchial artery; the gill lamallae of the gill pouches.

EFFERENT COLLECTOR LOOPS — They encircle each of the first four gill pouches. The fifth gill pouch has no demibranch on its posterior surface, thus there is no fifth collector loop, only four and a half. Each loop consists of a *pretrematic* branch on the anterior side of each gill pouch and a *post-trematic* branch on the posterior side. Adjacent collector loops are connected to one another by *intertrematic branches* which pass through the interbranchial septa.

Other Branches of the Efferent Branchial Arteries

The efferent branchial arteries give off many branches. These carry oxygenated blood to the more anterior parts of the shark's body.

EXTERNAL CAROTID ARTERY — Arises from the ventral end of *first collector loop*. This small vessel runs anteriorly along the lower jaw. Trace it as far as you can.

HYOIDEAN EPIBRANCHIAL ARTERY (EFFERENT HYOIDEAN ARTERY) — This artery arises from the dorsal end of the first collector loop. It runs anteriorly across the roof of the mouth, then divides in two, the *stapedial* and the *internal carotid* arteries.

The *stapedial artery* passes to the eye to supply some of its outer muscles. The *internal carotid artery* passes sharply to the midline along the ventral surface of the skull. The right and left branches unite temporarily, enter the cranial cavity, then separate and pass laterally to the brain. Soon after entering the chondrocranium, it is joined by the *efferent spiracular artery*.

AFFERENT SPIRACULAR ARTERY — This short vessel arises from the middle of the pretrematic artery of the first collector loop. It passes to the spiracle where it branches to supply blood to the lamellae of the pseudobranch. *Note:* the afferent spiracular artery carries *oxygenated* blood to the pseudobranch. Thus, the spiracular pseudobranch has no role in respiration.

EFFERENT SPIRACULAR ARTERY — This artery exits the spiracle, runs anteromedially into the chondrocranium where it joins the *internal carotid* artery. It turns ventrally past the *stapedial artery* to give rise to the *ophthalmic artery*.

OPHTHALMIC ARTERY — After arising on the *efferent spiracular artery* it passes anteriorly to supply blood to the eye.

PAIRED DORSAL AORTAE (RADIX AORTA) — They are anterior extensions of the unpaired *dorsal aorta*. These paired slender vessels originate from the first efferent branchial arteries near the mid-line. They pass anteriorly and near the level of the spiracles they turn laterally to join the *hyoidean epibranchial arteries*.

COMMISSURAL ARTERY (HYPOBRANCHIAL ARTERY) — This artery arises from the ventral end of the *second collector loop,* passes posteriorly to receive branches from the third collector loop. It continues ventral to the afferent branchial arteries to join and anastomase with its partner from the opposite side.

PERICARDIAL AND CORONARY ARTERIES — Upon entering the pericardial cavity, the commisural artery forms the *pericardial arteries* to the walls of the cavity, and the *coronary arteries* to the ventral surfaces of the conus arteriosus and the ventricle.

ESOPHAGEAL ARTERY (PHARYNGOESOPHOGEAL ARTERY) — Although most lower body parts are supplied by branches of the dorsal aorta, the roof of the pharynx and esophagus are supplied by the *esophageal artery* which originates from the second efferent branchial artery near its origin at the dorsal end of the collector loop. It extends posteriorly to the roof of the pharynx and esophagus.

THE DORSAL AORTA AND ITS BRANCHES

Oxygenated blood passes to most of the body by a branch of the *dorsal aorta*. It supplies both the body wall (epaxial and hypaxial muscles) and the visceral organs.

The Dissection

It will be necessary to view the origins of the aorta in the pharynx and its passage posteriorly to the tail. The dissection of the *pharynx* and the efferent branchial vessels has already been done. The *pleuroperitoneal cavity* has also been opened and the viscera examined. Organs will need to be moved and arterial branches followed with probes and dissecting needles.

Locate and identify the branches of the aorta which are named below. Trace each as far as you can.

DORSAL AORTA — Examine the previously dissected roof of the pharynx. Find the four pairs of *efferent branchial arteries.* They join at the dorsal midline to form the large arterial trunk, the *dorsal aorta,* which passes posteriorly bringing oxygenated blood from the gills to virtually every part of the shark's body, both its musculature and visceral organs. It passes through the *transverse septum* to enter the *pleuroperitoneal cavity*. There it continues posteriorly, giving off branches along its path, toward the tail. Some of the branches are paired, some are unpaired.

Subclavian Artery — This paired artery, the first major branch of the aorta, originates in the posterior pharynx. It supplies blood primarily to the pectoral fins and the muculature of the body wall. The *subclavian artery* originates from the dorsal aorta between the third and fourth efferent branchial arteries. Follow it as it curves ventrally toward the body wall where it gives off several branches.

Lateral Artery — This is the first of the branches of the subclavian artery. It runs parallel and follows the lateral canal posteriorly.

Ventrolateral Artery — This is a second branch, which also passes posteriorly to supply blood to the body musculature. It lies midway between the lateral line and the mid-ventral line.

Brachial Artery — This is the continuation of the subclavian artery passing into the pectoral fins.

The *dorsal aorta* continues posteriorly into the pleuroperitoneal cavity. Expose this cavity. In the mid-dorsal line, between the two kidneys, locate the dorsal aorta. Locate and identify all of the branches of the aorta named below. Study the accompanying photos to help you learn them.

The dorsal aorta gives off four unpaired arteries to the viscera:

1. **Celiac Artery**
2. **Anterior Mesenteric**
3. **Lienogastric**
4. **Posterior Mesenteric**

Celiac Artery — This is the most anterior branch of the dorsal aorta in the pleuroperitoneal cavity. It is also the most extensive; with repeated branches extending to almost all of the visceral organs. The artery passes ventrally to give off small branches to the gonads, the *genital arteries (ovarian arteries* in females, *spermatic arteries* in males). At times these may come directly from the aorta. Other small branches pass to the *esophagus* and the cardiac end of the *stomach*. It then continues posteriorly to give off two main branches: the *gastrohepatic artery* and the *pancreaticomesenteric artery*.

Gastrohepatic Artery — This extremely short artery branches almost at its origin to form the *gastric artery* and the *hepatic artery*.

The Gastric Artery — This artery, as the name implies, supplies the stomach. It divides into *dorsal* and *ventral* branches which supply the dorsal and ventral sides of the stomach.

The Hepatic Artery — This narrower branch turns sharply anteriorly running parallel to the bile duct to supply the liver.

The Pancreaticomesenteric Artery — This last branch is a continuation of the celiac artery posteriorly. It first runs behind the pylorus of the stomach to give off three branches, then continues posteriorly as the *anterior intestinal artery*. The first three branches are the *duodenal artery* to the duodenum, the *pyloric artery* to the pylorus of the stomach, and the *intraintestinal artery* which passes into the valvular intestine.

Anterior Intestinal Artery — This continuation of the pancreaticomesenteric artery runs down the right side of the valvular intestine.

Anterior Mesenteric Artery (Posterior Intestinal Artery) — This next branch of the aorta supplies the left side of the valvular intestine.

Lienogastric Artery (Gastrosplenic Artery) — This next branch of the dorsal aorta passes to the dorsal lobe of the pancreas, to the spleen and stomach.

Posterior Mesenteric Artery — The most posterior of the unpaired visceral branches of the aorta passes along the margin of the mesorectum to supply the rectal gland.

The dorsal aorta also gives off numerous paired somatic branches, the *parietal arteries*, to the body wall.

Renal Arteries are also given off to the kidneys. Beyond the point where the posterior mesenteric artery is given off, the dorsal aorta passes dorsal to the kidneys and is not seen without further dissection.

Iliac Arteries — In the vicinity of the cloaca the dorsal aorta gives off paired *iliac arteries,* which supply blood to the cloaca and the pelvic fins.

The Caudal Artery — This is the most posterior portion of the dorsal aorta; the continuation to the end of the tail. It can be seen in a cross-sectional view of the tail lying in the hemal arches of the vertebrae.

VENOUS SYSTEM

The Hepatic Portal System

The venous system is involved in the return of blood to the heart. In the shark, besides *veins* and smaller *venules*, one finds *venous sinuses* which are large, thin-walled spaces for the collection of venous blood. These are usually injected with *blue latex*.

A *portal system* is a venous system which begins as capillaries in an organ and ends as capillaries in another organ.

The *hepatic portal system* begins as capillaries in the organs of the digestive system. These join together to form the large vein, the *hepatic portal vein*. This vein then enters the liver where it subdivides again to form capillary-like tubes and venous sinuses. Within the liver the products of the digestive organs undergo metabolic processing. The blood from the liver is collected into the hepatic veins (sinuses) to empty into the *sinus venosus* of the *heart*.

In triply injected specimens the hepatic portal system has been injected with yellow latex.

We are choosing this venous system first because we have just examined the arteries leading to organs of the digestive system, and veins of the hepatic portal generally travel together with those arteries. Thus, no further dissection need be made to view this system.

THE HEPATIC PORTAL VEIN — This thick vein may be found in the *lesser omentum,* a mesenteric membrane also known as the *gastro-hepato-duodenal ligament*. It lies alongside and somewhat dorsal to the *bile duct*. It receives small *choledochal veins* from the bile duct. It is, however, formed more posteriorly by the joining of three large visceral veins:

1. **Gastric Vein**
2. **Lienomesenteric Vein**
3. **Pancreaticomesenteric Vein**

Gastric Vein — This is the most anterior of the three branches. Trace this vein to the stomach. Note that it is formed in the middle of the stomach from two *dorsal* and two *ventral gastric veins*. Also note that the *gastric vein* and *artery* travel together.

Lienomesenteric Vein — Posteriorly this vein originates from two others: the *posterior intestinal vein* and the *posterior lienogastric vein*.

> **Posterior Intestinal Vein** — This vessel can be seen coming from the left posterior surface of the valvular intestine. Note the transverse *annular veins* which indicate the lines of attachment of the spiral valve.
>
> **Posterior Lienogastric** — This vein comes from the spleen and the posterior end of the stomach.
>
> The unified *lienomesenteric vein* passes anteriorly from the spleen toward the duodenum along the dorsal surface of the pancreas.

Pancreaticomesenteric Vein — This is the right branch of the hepatic portal vein. It may be seen near the ventral lobe of the pancreas at the duodenum. There it is joined by a small vein, the pyloric vein, which comes from the pylorus of the stomach. It also receives the *intraintestinal vein* from within the spiral valve. Note that these veins travel together with the arteries of the same name. Near the pyloric vein it also receives the *anterior lienogastric vein (anterior splenic vein)* from the spleen. It lies along the pyloric region of the stomach.

Anterior Intestinal Vein — The most posterior portion of the pancreaticomesenteric vein ascends along the right side of the valvular intestine collecting blood from *annular veins* along its way.

The Renal Portal System

The renal portal system is the second venous portal system in the shark. It begins as capillaries in the tail and ends as capillaries in the *kidneys*. Systemic veins (not part of the renal portal system) then return the blood from the kidneys to the heart. The veins of this system have not been injected with a uniquely colored dye and it may be difficult to trace the smaller vessels.

KIDNEYS — You will recall from your examination of the pleuroperitoneal cavity that the kidneys are located along the dorsal body wall, on either side of the mid-line, as two elongate, ribbon-like structures.

CAUDAL VEIN — Recall this vein from our study of the muscular system when we viewed a cross-section of the tail; see photo on page 41. In the ventral portion of the vertebra, below the centrum, locate the hemal arch, carrying blood anteriorly from the tail. By making several more cross-sectional cuts every half inch and proceeding anteriorly you will find that the *caudal vein* bifurcates and proceeds anteriorly as two veins.

RENAL PORTAL VEINS — These are the two veins which result from the bifurcation of the caudal vein. They proceed anteriorly, lying dorsal to and in the lateral border of the kidneys. Lift up the dark ribbon-like kidney at its lateral border. Examine it dorsally with a hand lens.

AFFERENT RENAL VEINS — Lift the dark ribbon-like kidney at its lateral border. Examine its dorsal surface with a hand lens. The numerous small side branches that empty into the kidney are the *afferent renal veins*.

EFFERENT RENAL VEINS — Blood from the kidneys passes through other small and numerous vessels located dorso-medially, the *efferent renal veins*. It enters into elongated *posterior cardinal veins* which pass on either side of the mid-dorsal line. They return the blood from the kidneys to the heart.

The Systemic Veins

The systemic veins of the shark are all those returning blood to the heart that are not a part of the venous portal systems. They are generally paired.

The veins are, in most cases, dilated tissue spaces, without clearly defined walls. They are more properly called *sinuses*. This lack of definite structure makes identification more difficult. It will be helpful if your specimen is injected. Generally the systemic veins are injected with *blue latex*. The latex will conform to the original shape of the vein permitting one to follow even the narrower veins.

We shall begin our study of the systemic veins with the *sinus venosus* of the heart.

SINUS VENOSUS — Eventually all of the systemic veins return blood to the most posterior chamber of the heart, the *sinus venosus*. This chamber was already studied when the heart was discussed; see page 59. Two pairs of veins enter the sinus venosus:

1. **Hepatic Veins**
2. **Common Cardinal Veins (Ducts of Cuvier)**

HEPATIC VEINS (SINUSES) — Probe the posterior wall of the sinus venosus. Find two slits just lateral to the midline. Follow the *hepatic veins* through the transverse septum into the hepatic sinuses of the liver.

COMMON CARDINAL VEINS (DUCTS OF CUVIER) — These vessels enter the lateral corners of the sinus venosus, one on each side. They collect blood from four major branches: the *anterior cardinal veins*, the *inferior jugular veins*, the *posterior cardinal sinuses*, and the *subclavian veins*.

Anterior Cardinal Veins — These veins drain the brain and all of the head except the floor of the branchial region. Anteriorly the *anterior cardinal veins* dilate to form the *anterior cardinal sinuses*. These may have been viewed when studying the branchial musculature. They lie laterally, dorsal to the gill arches, beneath the cucullaris muscles, which must first be removed. The anterior cardinal sinuses originate in the brain.

Orbital Sinuses — In the region of the eye the anterior cardinal sinuses communicate with the *orbital sinuses* which surround the eye. Below the orbits, the *interorbital sinus* connects the right and left anterior cardinal sinuses. Probe this passageway with a flexible probe.

Inferior Jugular Veins — Another set of veins draining the head area are the *inferior jugular veins*. They drain the floor of the mouth and the ventral gill area. In their passage posteriorly they are joined to the anterior cardinal sinuses by a short *hyoidean sinus*, one on each side of the head. It lies posterior to the hyoid arch. Trace the inferior jugular veins posteriorly until they enter the anterior surfaces of the *common cardinal veins*.

Posterior Cardinal Sinuses — From the posterior parts of the body the common cardinal veins receive the paired *posterior cardinal sinus*. It is a large, bluish, thin-walled chamber at the anterior end of the pleuroperitoneal cavity toward the mid-dorsal line. The right and left posterior cardinal sinuses join mid-dorsally at the level of the gonads.

Genital Sinuses — At the level of the gonads the posterior cardinal sinuses receive genital blood vessels (ovarian or testicular veins) from *genital sinuses* that lie beside the gonads. They also receive vessels from the *esophagus*.

Posterior Cardinal Veins — Caudally, the posterior cardinal sinuses continue as two narrow veins, the *posterior cardinal veins*. These run along the dorso-medial line of the pleuroperitoneal cavity. They lie dorsal and medial to the kidneys, and lateral to the dorsal aorta. The right posterior cardinal vein extends further caudally than the left one. They receive blood from the kidneys by way of the tiny numerous *efferent renal veins*.

Subclavian Veins — These short veins enter the *common cardinal veins* just lateral to the entrance of the *inferior jugular veins*. They are formed by the union of three veins: the *brachial veins*, the *subscapular veins*, and the *lateral abdominal vein*.

Lateral Abdominal Veins — These veins are readily seen as dark longitudinal lines when inspecting the lateral walls of the pleuroperitoneal cavity, right beneath the parietal peritoneum. They drain the fins and the lateroventral trunk musculature. They also drain the *cloacal veins* which come from the lateral wall of the cloaca. In addition, blood from the pelvic fins, drained by way of the *femoral veins*, continues anteriorly into the *iliac veins* before emptying into the *lateral abdominal veins*. Anteriorly, at the level of the pectoral girdle they receive the *brachial veins* which drain the pectoral fins.

Brachial Veins — Anteriorly, at the level of the pectoral girdle, the *brachial veins* which drain the pectoral fins join the lateral abdominal veins, near their entry into the subclavian veins.

Subscapular Veins — From their dorsal origins in the area of the pelvic girdle, the *subscapular veins* pass ventrally to join the subclavian veins.

Coronary Veins — These may be seen on the surface of the ventricle of the heart. They enter the sinus venosus by a common aperture.

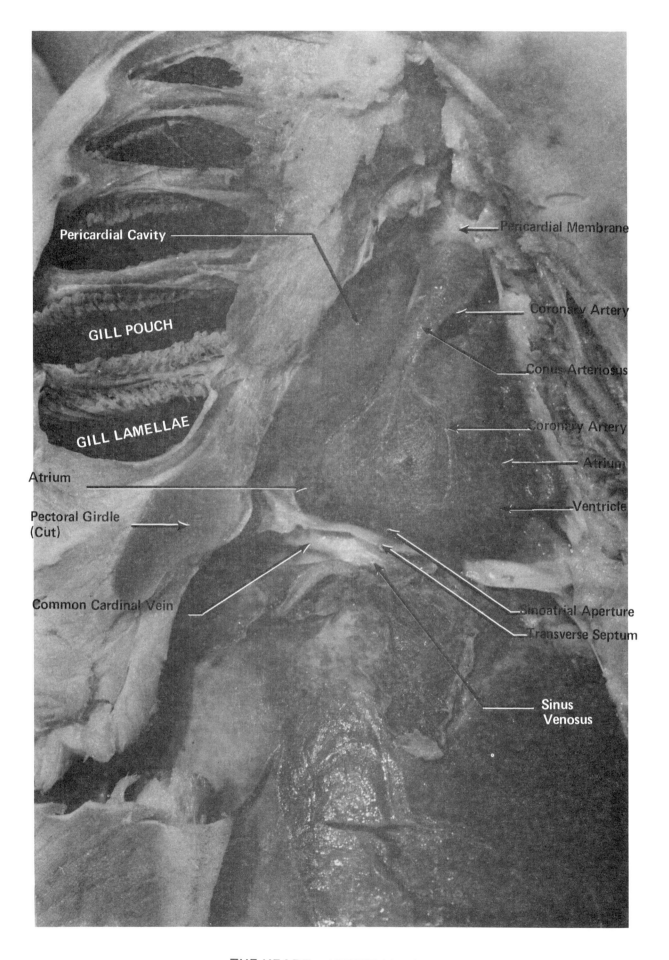

THE HEART — VENTRAL VIEW

THE BRANCHIAL ARTERIES

THE BRACHIAL ARTERIES (CLOSE-UP)

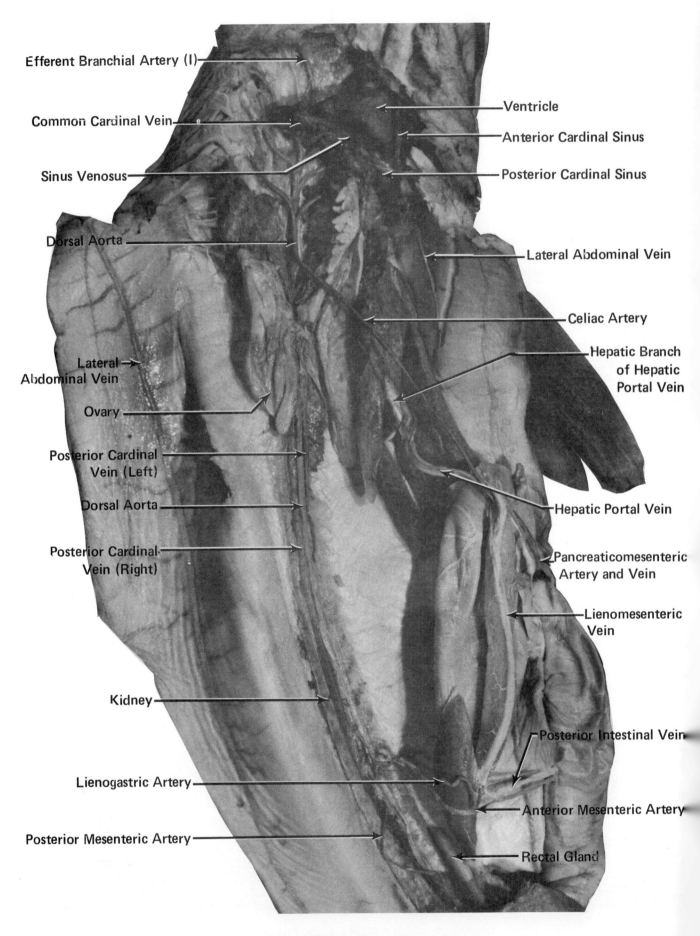

MAJOR BLOOD VESSELS OF TRUNK

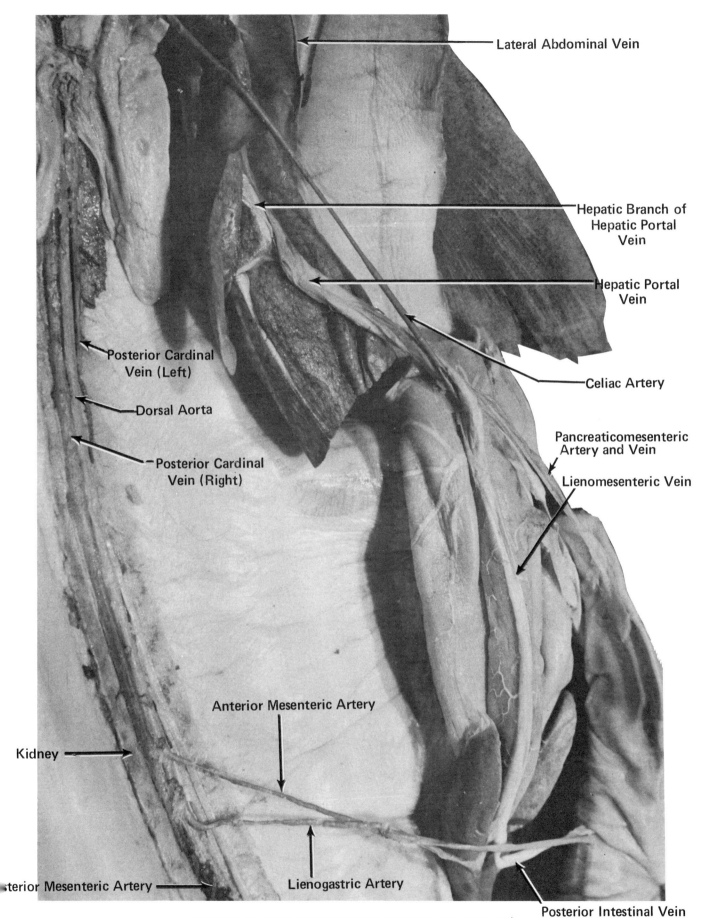

MAJOR BLOOD VESSELS OF TRUNK (CLOSE-UP)

71

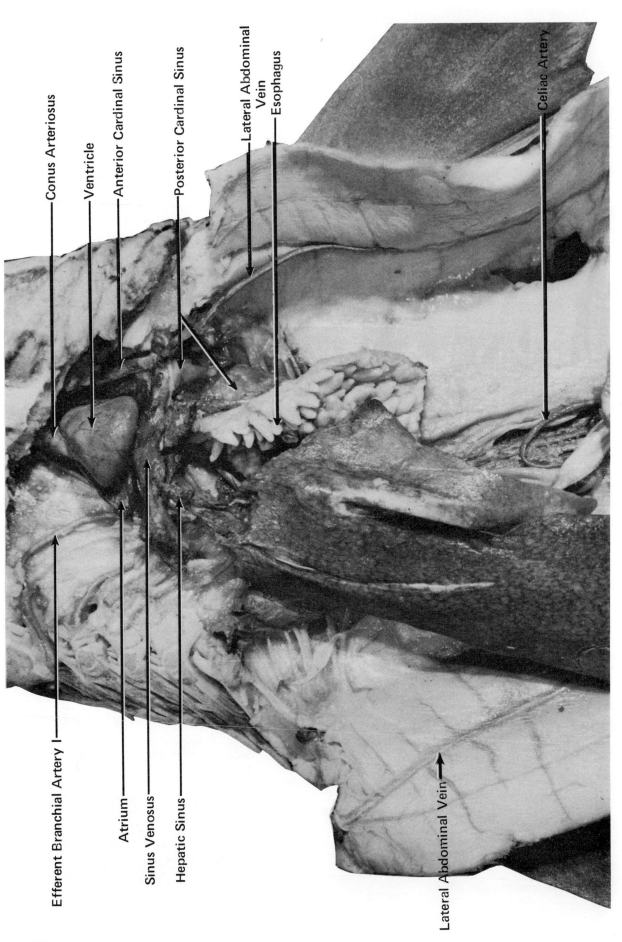

SYSTEMIC VEINS

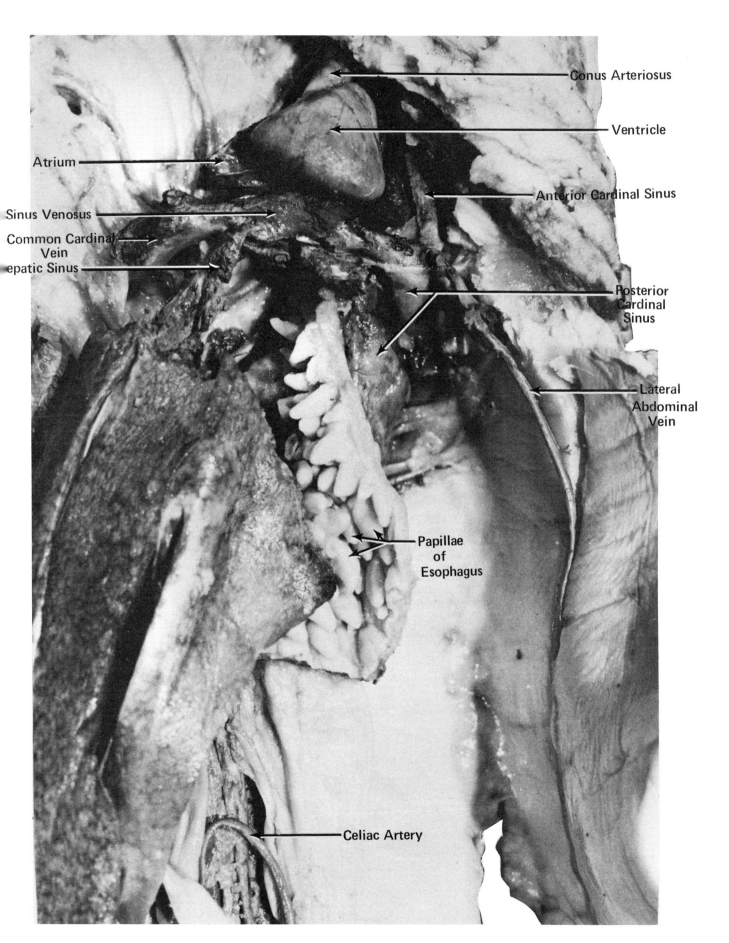

SYSTEMIC VEINS (CLOSE-UP)

Name _____ Section _____ Date _____

SELF-QUIZ IV
THE CIRCULATORY SYSTEM

1. Describe the path of the blood through the heart.
2. How is the kidney supplied with blood? How is it drained?
3. Describe the path of blood from the heart to the gills.
4. Describe the circulation of blood at the gills.
5. Describe the path of oxygenated blood from the gills to the trunk.
6. How do the blood sinuses differ from veins?
7. What are the major branches of the dorsal aorta?
8. What are the major branches of the celiac artery?
9. Name the major veins drained by the hepatic portal vein.
10. Describe the function of each of the structures listed below.

ANSWERS

1. _____
2. _____
3. _____
4. _____
5. _____
6. _____
7. _____
8. _____
9. _____
10. a. sinus venosus _____
    b. conus arteriosus _____
    c. intertrematic vessels _____
    d. afferent branchial arteries _____
    e. efferent branchial arteries _____
    f. anterior cardinal sinus _____
    g. inferior jugular vein _____
    h. posterior cardinal sinus _____
    i. orbital sinus _____
    j. afferent renal vein _____

Label all of the features indicated on the following illustration.

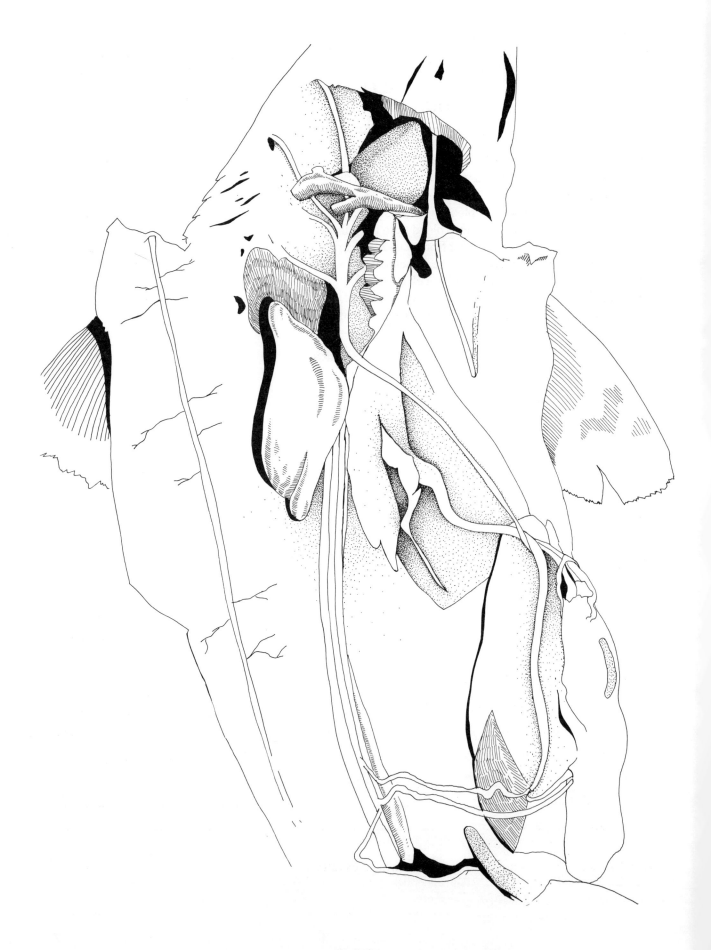

MAJOR BLOOD VESSELS OF THE TRUNK

THE UROGENITAL SYSTEM

The urinary and genital systems have distinct and unique functions. The first, the removal of nitrogenous wastes and the maintenance of water balance; the other, the reproduction of the species. However, due to their similar developmental origins and the sharing of common structures, they are usually considered as a single system.

The shark kidney and its ducts are quite different from those in higher vertebrates. The relationship between the urinary and genital structures is also quite different. Male and female sharks differ in their urinary as well as in their genital systems. The urogenital system of the shark illustrates a simpler stage of development.

The Dissection

Mature specimens make for the best dissections since the entire urogenital system will be fully developed. In immature specimens most structures are undifferentiated. If your animal is a male, you are also responsible for studying and knowing the urogenital system of an adult female shark, and vice versa.

Expose the pleuroperitoneal cavity. Remove almost the entire liver except for its anterior end. Cut the esophagus about a half inch from its entry into the body cavity. Then cut the colon about one and a half inches from its posterior end. Free the alimentary canal, pancreas, and spleen from their mesentery and vascular connections and remove entirely from the body. This will reveal the urogenital structures: gonads, kidneys, and associated ducts.

Further dissection in this chapter is minor and will be indicated in the text as each is approached.

We shall study the female first, then the male. This will be followed by a discussion of fertilization and development in the dogfish shark.

FEMALE

KIDNEYS — The *kidneys* are flattened, ribbon-like, darkly colored structures lying dorsally on either side of the midline, along the entire length of the pleuroperitoneal cavity. They lie *retroperitoneally*, behind the transparent peritoneum. A tough white glistening strip of connective tissue, the *caudal (innominate) ligament* is found between the kidneys in the midline.

In females, the upper portion of the kidney is nonfunctional; the formation of urine and the removal of wastes take place in the lower portion.

ARCHINEPHRIC DUCT (OPISTHONEPHRIC DUCT) (WOLFFIAN DUCT) — These urinary ducts are narrow in the female and difficult to see. They extend posteriorly along the ventral surface of each kidney. Distally, near the cloaca, the urinary ducts widen to form the *urinary vesicles,* then open by pores into the *urinary papilla*. In males, these ducts lose their urine transporting function and transport seminal fluids and sperm.

URINARY PAPILLA — This fleshy conical projection is readily seen emerging from the cloaca. It is the exit point for urine to the external environment.

OVARIES — Look within the anterior part of the pleuroperitoneal cavity, dorsal to the liver. Locate two cream-colored elongated organs on either side of the mid-dorsal line. Each *ovary* is supported by a mesenteric membrane known as the *mesovarium*. The shape of the ovaries will vary depending upon the maturity of the specimen. In immature females they will be undifferentiated and glandular in appearance. In mature specimens you may find two to three large eggs, about three centimeters in diameter, in each ovary. When these break the surface of the ovary, upon *ovulation*, they enter the body cavity and by means of peritoneal *cilia* are moved into the oviducts.

OVIDUCTS — The oviducts are elongated tube-like structures lying dorsolaterally the length of the pleuroperitoneal cavity, along the sides of the kidneys. In mature specimens they are more prominent, each suspended dorsally by a *mesotubarium*, a mesenteric membrane extending from the kidney. The distal half of the oviduct is enlarged to form the *uterus*. Trace one oviduct anteriorly. It passes dorsal to the ovary, then curves ventrally in front of the anterior portion of the liver. The oviducts from opposite sides unite near the anteroventral border of the liver. Here the *ostium* may be found.

Ostium — This is the single common opening of both oviducts. It is located within the *falciform ligament*, the membrane that connects the anteroventral portion of the liver to the parietal peritoneum. The *ostium* is located in the posterodorsal side of the ligament, and can be opened by spreading apart its edges. In immature specimens the ostium cannot readily be opened.

Shell Gland (Nidamental Gland) — Near the anterior end of the oviduct locate an enlarged short segment of the duct known as the *shell gland*. It serves a double purpose. As the eggs pass the gland it secretes a thin horny shell around a group of several eggs. The mass, then known as the *candle*, passes down the oviduct. The second function of the shell gland is to serve as a reservoir for the storage of sperm. Thus the eggs are fertilized and receive a light shell-like covering as they pass through the shell gland.

UTERUS — The posterior half of the oviduct becomes enlarged and is known as the *uterus*. Here the fertilized eggs develop into embryos. Upon completing their period of gestation (close to two years) the young are ready to be born. A fuller description of fertilization and development in the dogfish shark is given toward the end of this chapter.

CLOACA — This opening serves for the elimination of urinary and fecal wastes as well as an aperture through which the young "pups" are born. The two uteri open into the posterodorsal portion of the cloaca just ventral to the urinary papilla. The urogenital portion of the cloaca is known as the *urodeum* and is partly separated by means of horizontal lateral folds from the more anteroventral portion, where the rectum terminates, the *coprodeum*.

SUPRARENAL BODIES (ADRENAL GLANDS) — Although not readily located in the shark as distinct structures, they are identified as a series of pale spots found longitudinally upon the medial surface of the kidney, near the dorsal midline. Staining and microscopic examination are needed to verify their glandular nature.

MALE

KIDNEYS — The kidneys of the male are essentially the same as those just described in the female. The posterior portion is involved in the manufacture and transport of urine, its role quite similar to that in females. The main difference lies in the anterior portion of the kidney, which in females is degenerate and functionless, but in males is an active part of the reproductive system.

TESTES — Paired *testes* lie near the anterior end of the pleuroperitoneal cavity, dorsal to the liver, adjacent to the anterior ends of the kidneys. Each testis is supported by a mesenteric membrane known as a *mesorchium*. It is across this membrane that sperm pass from the testes to the kidneys within narrow tubules called *efferent ductules*. They are too small to be seen without a hand lens.

EPIDIDYMIS — Sperm pass from the efferent ductules to the anterior ends of the kidneys. This portion of the kidney is known as the *epididymis*. It has virtually no urinary function.

DUCTUS DEFERENS (ARCHINEPHRIC DUCT) (WOFFIAN DUCT) — After passing through the epididymis the sperm enter the *ductus deferens* and pass posteriorly toward the cloaca. In mature male specimens the ductus deferens may be seen on the ventral surface of the kidneys as a pair of highly coiled tubules. The kidney right below the epididymis is known as *Leydig's gland*. Here the secretion from the testes is modified as a milky thick fluid analogous to the seminal fluid of higher vertebrates. This duct is also known as the Wolffian duct.

Note: While in the female this duct carries urine, in the male it transports spermatozoa and seminal fluid.

SEMINAL VESICLE — The posterior portion of the ductus deferens widens and straightens to form the paired *seminal vesicles*. Nick the surface of one with a pin and observe a thick white fluid oozing out. This is the *seminal fluid*.

SPERM SACS — These paired sacs at the posterior ends of the seminal vesicles receive the seminal secretions. They join to form the *urogenital sinuses* which exit through the fleshy conical *urogenital papilla* which may be seen extending from the cloaca.

ACCESSORY URINARY DUCTS — Since the archinephric (Wolffian) ducts in males collect and transport seminal fluid, another set of tubes, the *accessory urinary ducts*, collect and transport urine from the kidneys. These paired thin tubules may be found along the medial side of the posterior half of the kidney. It is necessary to push aside the seminal vesicles in order to see the urinary ducts. Small collecting tubules from the kidneys lead into the accessory urinary ducts along their lengths. They exit through the *urogenital sinus* and *urogenital papilla* just as the sperm sacs did.

UROGENITAL SINUS and **UROGENITAL PAPILLA** — As we have seen, the *urogenital sinus* serves as the receptacle for the seminal vesicle and for the accessory urinary ducts. The *urogenital papilla* also serves both urinary and genital systems. Thus, their names are well-deserved.

CLOACA — As in the female, this structure receives the rectal wastes as well as the genital and urinary products. As in the female, the cloaca is divided by horizontal lateral folds to form a *urodeum* portion where the urogenital products empty and a more anteroventral portion, the *coprodeum,* where the rectum terminates.

Two additional male structures will be studied. They have no homologues in females.

SIPHON — Make a transverse cut into the ventral surface of one pelvic fin. Find a thin-walled muscular sac, the *siphon*. It is a closed blind sac anteriorly, posteriorly it is connected to the *dorsal groove* of the *clasper*.

CLASPERS — They are modified extensions of the medial portions of the pelvic fins. They are inserted into the female's cloaca during copulation.

Clasper Tube — The finger-like claspers each have a dorsal groove, the *clasper tube (spermatic sulcus)* that carries the seminal fluid from the cloaca of the male to the cloaca of the female during mating.

It was originally thought that the *siphon sac* was filled with sea water, which during copulation was ejected along the *clasper tube* to help propel the sperm toward the female. It was subsequently shown, however, that the siphon sacs secrete large amounts of mucopolysaccharide which may lubricate the claspers and contribute to the seminal fluid.

FERTILIZATION AND DEVELOPMENT

We have already pointed out that fertilization in the dogfish shark is *internal,* usually taking place within the *shell gland* of the *oviduct*. The fertilized eggs continue to move posteriorly toward the *uterus*. Here the young develop.

As they grow they are attached to the egg, now known as the *yolk sac,* by means of a stalk.

Dissect a "pup" of about 15 centimeters in length and note an *internal yolk sac,* continuous with the external yolk sac, connected to the alimentary canal.

During its period of gestation, which is nearly two years, the yolk is slowly absorbed by the shark "pup." At about 25 centimeters in length the external yolk sac has been completely absorbed, although some of the internal yolk sac is still present. At birth the young are about 23 to 29 centimeters long.

Numerous uterine *villi,* finger-like projections from the uterine wall, make contact with the surface of the developing embryo and its yolk sac. It is believed that these provide the embryo with water; all other nutrients are supplied by the yolk.

This type of development, where the young are born as miniature adults but have received hardly any nutrition directly from the mother's uterus, is known as *ovoviviparous*. By contrast, human development is *viviparous*.

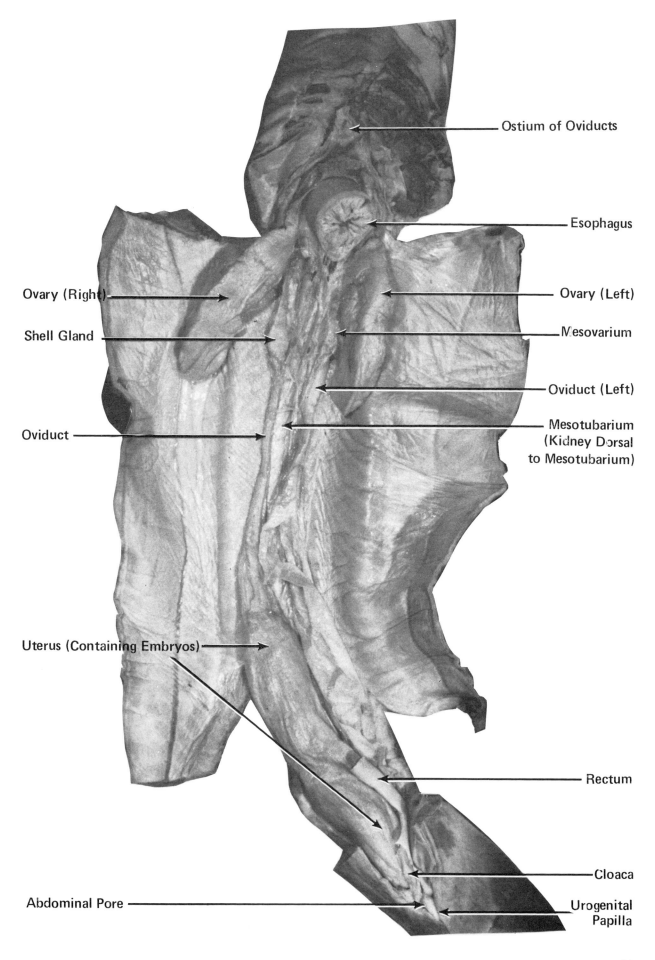

THE FEMALE UROGENITAL SYSTEM

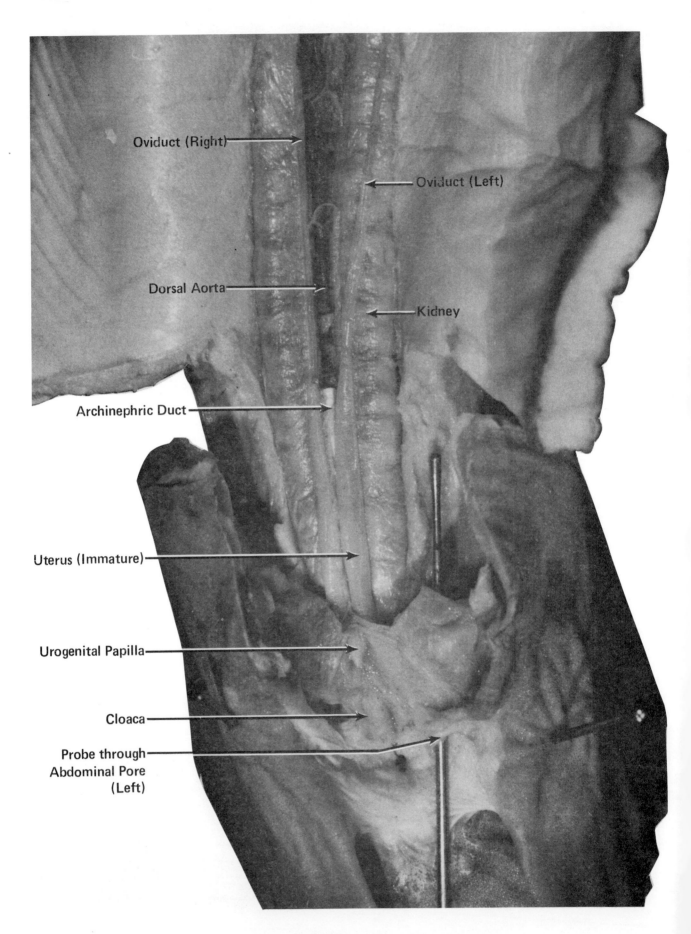

THE FEMALE UROGENITAL SYSTEM (CLOSE-UP)

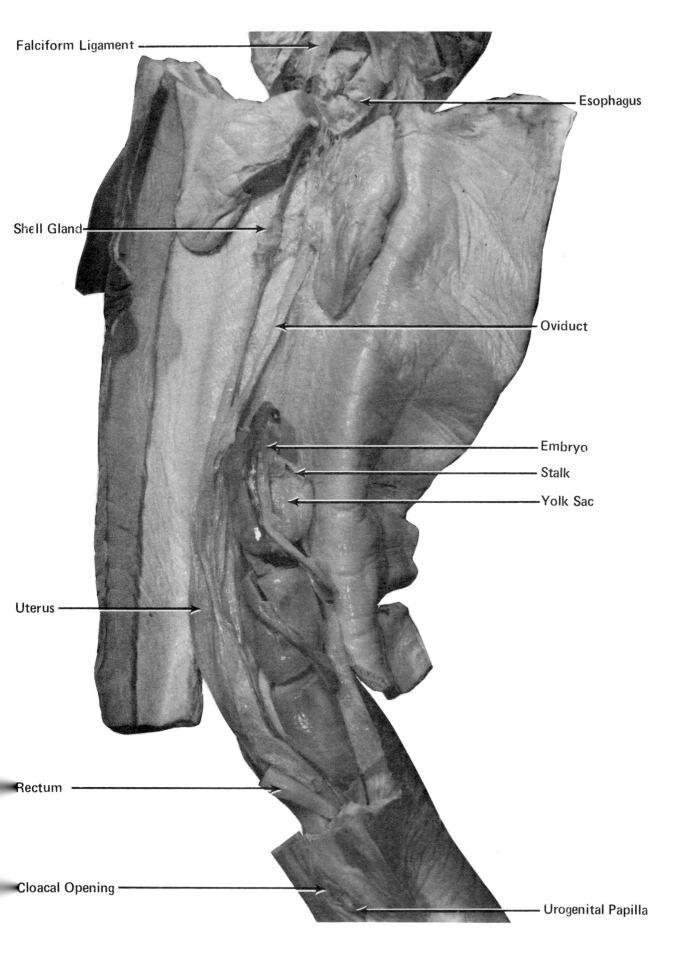

EMBRYOS WITHIN UTERUS

EMBRYOS REMOVED FROM UTERUS

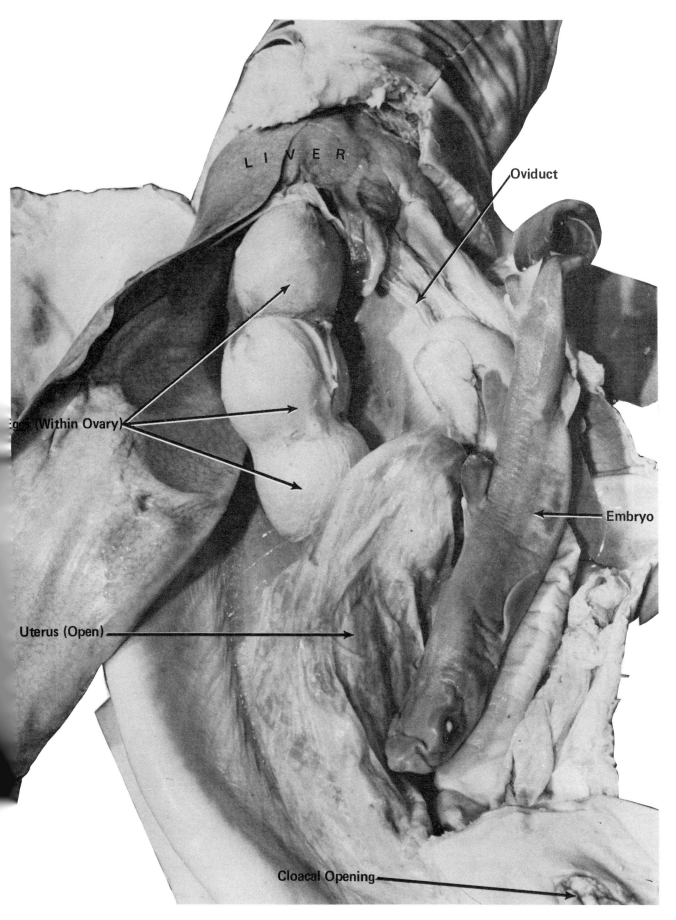

MATURE EMBRYO IN UTERUS

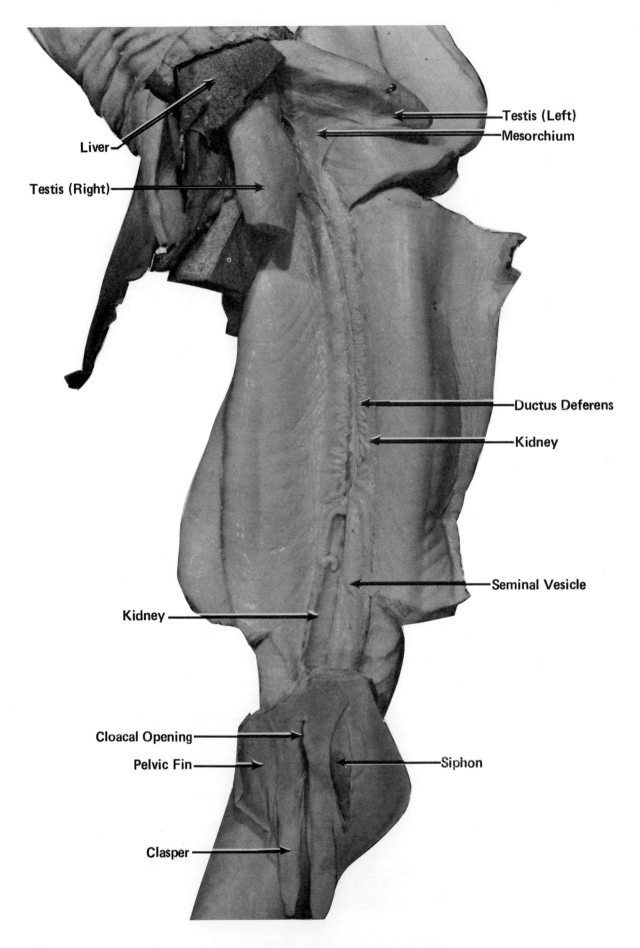

THE MALE UROGENITAL SYSTEM

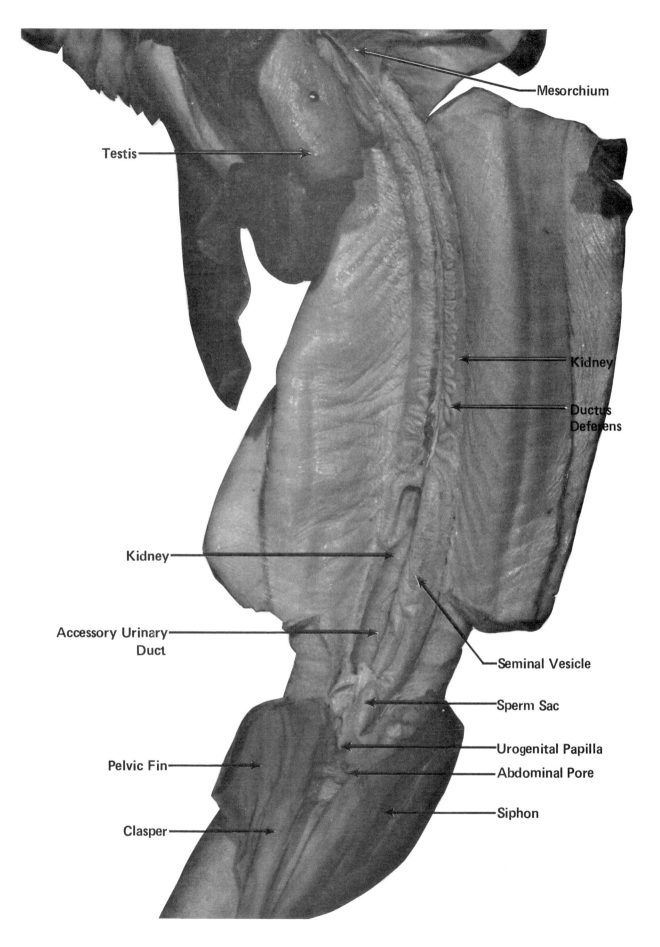

THE MALE UROGENITAL SYSTEM (CLOACA EXPOSED)

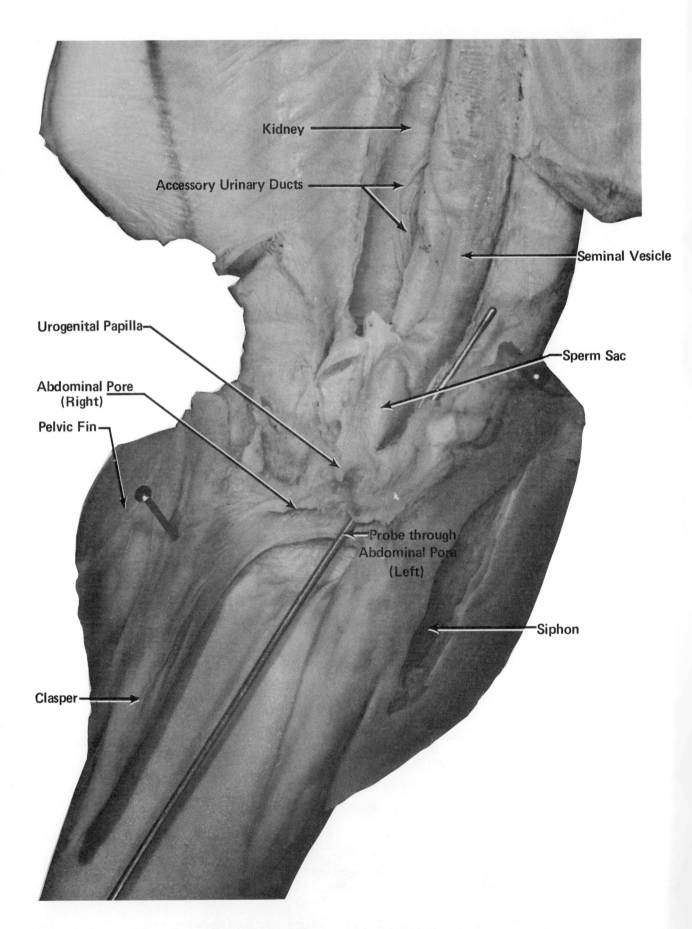

THE MALE UROGENITAL SYSTEM (CLOSE-UP)

Name _____ Section _____ Date _____

SELF-QUIZ V
THE UROGENITAL SYSTEM

1. Describe the transport of urine in the male and in the female shark.
2. Describe the path of the egg from the ovary to the cloaca.
3. Describe the path of the sperm from the testis to the cloaca.
4. Describe the differences between the various stages of development of fetal sharks as found in the uterus.
5. Describe the relationship between the lower urinary, genital, and digestive tracts as they are found within the cloaca of female sharks.
6. The same question as the previous one, but for males.
7. How is copulation accomplished in sharks?
8. How does the kidney of sharks differ from the human kidney?
9. What is the relationship between the abdominal pores and the urogenital system?
10. Define each of the terms listed below.

ANSWERS

1. _____
2. _____
3. _____
4. _____
5. _____
6. _____
7. _____
8. _____
9. _____
10. a. retroperitoneal _____
    b. seminal vesicle _____
    c. Wolffian duct _____
    d. mesorchium _____
    e. shell gland _____
    f. gestation _____
    g. siphon _____
    h. accessory urinary tract _____
    i. ostium _____
    j. mesotubarium _____

Label all of the features indicated on the following illustration.

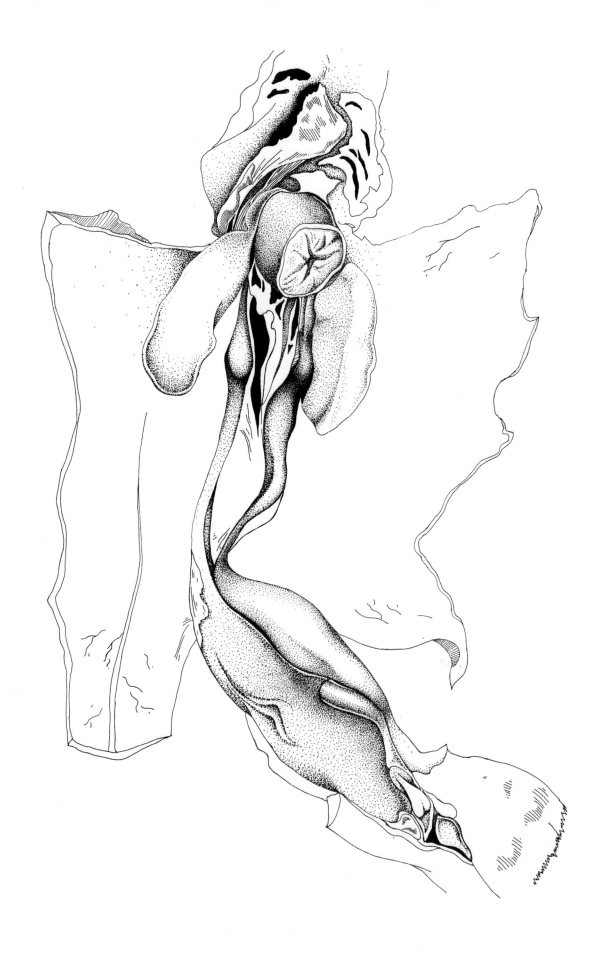

THE FEMALE UROGENITAL SYSTEM

THE NERVOUS SYSTEM: THE BRAIN AND SPINAL CORD

The nervous system functions in *communication* between the various parts of an organism and between the organism and its external environment. It consists of the *central nervous system;* the brain and spinal cord, and the *peripheral nervous system*; the sense organs, cranial and spinal nerves, and their branches.

We shall begin our study of the nervous system with an examination of the brain and the cranial nerves.

THE BRAIN

The advantages in studying the shark's brain include:

— The skull is composed of cartilage, not bone. This makes the brain accessible to the scalpel's blade.

— The brain of the shark is relatively large. Thus, the gross anatomy of smaller nerves and structures may be observed.

— It illustrates a lower level of development among vertebrates. Comparative studies of different vertebrates reveal ever-increasing complexity in the structure of the brain, especially in the cerebral region. The basic architectural plan of the vertebrate brain and cranial nerves is, however, already laid out.

The Dissection

The larger mature dogfish sharks make the best specimens for dissecting the brain since their structures and nerve tracts can be more readily observed than those of smaller specimens.

We shall dissect the left side of the brain first. In the process, the structures within the *otic capsule,* the left semi-circular canals of the *inner ear,* will be destroyed and underlying parts of the brain and cranial nerves revealed. Then, the right side of the brain will be dissected. This time the delicate inner ear and associated structures will be preserved, carefully dissected, and studied.

The technique of dissection of the chondrocranium is unique to cartilaginous fish, for unlike bone, the cartilage permits the use of a scalpel in exposing the brain. First remove the skin from the dorsal surface of the head from the rostrum posteriorly to the first gill slit. Continue removing the skin ventrally to the level of the eye and the spiracle.

Make all of your cuts of the chondrocranium horizontal and shallow in a shaving motion. This is your best guarantee that you will not be injuring delicate brain tissue or cranial nerve fibers. The cartilage is transparent up to a depth of about one millimeter. Therefore, cut very thin horizontal chips of cranium no more than one millimeter thick. The thin chips may be broken loose and removed with fine-toothed forceps. It is also helpful, once the outer layers of cartilage have been removed and the fine work of separating cartilage from nerve tissue begins, to use a smaller, No. 15 scalpel. This can be inserted into small spaces without causing the damage which might result from the use of a larger average-sized blade. The important rules here are:

—Cut only very thin chips, and
—Look carefully before you cut.

Your scalpel blades are extremely sharp and the soft delicate nerve tissue, unlike the cartilage, offers virtually no resistance.

Begin the careful removal of the cranium on the left side antero-dorsally and work your way posteriorly. As was pointed out, the inner ear on this side will be destroyed in order to see the brain and cranial nerves lying beneath it.

Study the photographs and diagrams in this section to help you locate and identify the parts of the brain and cranial nerves.

At this point we should become familiar with some of the terms used to describe the major subdivisions of the vertebrate brain:

- **Prosencephalon (Forebrain)**
  - **Telencephalon**
  - **Diencephalon**
- **Mesencephalon (Midbrain)**
- **Rhombencephalon (Hindbrain)**
  - **Metencephalon**
  - **Myelencephalon**

PRIMITIVE MENINX — As you expose the brain and spinal cord, you will find them enclosed within a delicate vascular protective membrane called the *primitive meninx*. It sends blood vessels to the surface of the brain and is also connected to the inner membrane of the cranial walls by fine strands of tissue. Examine it by removing some from the surface of the brain. In life, cerebrospinal fluid fills the space between the brain and the wall of the cranial cavity.

Prosencephalon (Forebrain)

TELENCEPHALON — This is the most anterior portion of the brain. Locate and identify each of the following:

Olfactory Bulbs — These are paired extensions of the anterior portion of the brain. They are rounded masses which make contact anteriorly with the spherical *olfactory sacs*, the organs of smell.

Olfactory Tracts — These are the long narrow nerve pathways extending posteriorly from the olfactory bulbs.

Cerebral Hemispheres (Cerebrum) — These two hemispheres are the rounded lobes of the anterior brain. Each lobe contains a cavity, the first and second ventricles of the brain, termed the *lateral ventricles*. These ventricles are continuous with the cavities in other parts of the brain, particularly the third and fourth ventricles. Trace the lateral ventricles to the central cavity in the olfactory tract and bulb. The anterior portions of the cerebrum are known as the *olfactory lobes*.

Olfactory Lobes — These are identified as the rounded masses at the anterior end of the cerebral hemispheres where the olfactory tracts terminate. They are separated from the rest of the cerebrum by externally visible indentations.

DIENCEPHALON — This is the second portion of the forebrain, directly posterior to the telencephalon. It is somewhat depressed with a dark-colored membrane covering a cavity, the *third ventricle*. The diencephalon is the part of the brain immediately surrounding the third ventricle.

Third Ventricle — This is an unpaired cavity in the diencephalon. It communicates anteriorly with the two lateral ventricles through an opening in the *foramen of Monro*. Posteriorly the third ventricle communicates with the fourth ventricle through a passageway, the *cerebral aqueduct (aqueduct of Sylvius)*.

The diencephalon consists of three parts:

1. Epithalamus — The area dorsal to the third ventricle. It includes:

Tela Choroidea — A thin roof consisting of a vascular membrane covers the diencephalon. It is rich in blood vessels and sends projections into the third ventricle to form the *anterior choroid plexus*, a delicate vascular membrane which secretes *cerebrospinal fluid* into the ventricles. Thus we see that the cerebrospinal fluid not only bathes the outside of the brain but is also the fluid of the ventricles and other cavities within the brain. The anterior portion of the roof forms a sac called the *paraphysis*.

Epiphysis (Pineal Body) — A slender stalk that projects anterodorsally from the rear of the diencephalon, an area known as the *habenula*, upwards through the roof of the chondrocranium by way of the *epiphysial foramen*. In most dissections it is destroyed when the roof of the chondrocranium is removed. Behind the pineal body is the *posterior commisure*, a tract of nerve fibers connecting the right and left sides of the brain.

2. Thalamus — An area of gray matter in the lateral walls of the third ventricle. These are best seen in a sagittal section of the brain.

3. Hypothalamus — It lies ventral to the third ventricle and is best seen in a ventral veiw of the brain. It forms the floor of the diencephalon and consists of the *infundibulum* and *hypophysis (pituitary body)* along the midventral line. Also seen in ventral view in this area is the optic chiasma, the point at which the optic nerves, one from each eye, cross medially and their fibers enter the opposite sides of the brain.

Mesencephalon (Midbrain)

OPTIC LOBES (OPTIC TECTUM) — Directly posterior and slightly dorsal to the diencephalon find a pair of prominent bulged structures. These are the *optic lobes*. They form the dorsal and lateral walls of the *mesencephalon*. Each optic lobe contains an *optic ventricle* which communicates ventrally with the *cerebral aqueduct (Aqueduct of Sylvius)*, the central cavity of the mesencephalon. The floor of the mesencephalon, the *tegmentum*, lies dorsal to the hypophysis.

Rhombencephalon (Hindbrain)

METENCEPHALON — Immediately posterior and somewhat dorsal to the optic lobes is the *cerebellum*.

Cerebellum — The oval-shaped dorsal portion of the metencephalon. It partly overlaps the optic lobes. Its inner cavity is known as the *cerebellar ventricle,* which communicates anteriorly with the cerebral aqueduct and posteriorly with the fourth ventricle. Externally, one may see two grooves, one longitudinal and one transverse, in the form of a cross, dividing the cerebellum into four sections. At the posterior end find two lateral ear-like projections of convoluted tissue known as *auricles of the cerebellum*.

MYELENCEPHALON — It lies posterior to the metencephalon and forms a major part of the *medulla oblongata*.

Medulla Oblongata — The elongated posterior region of the brain that is continuous posteriorly with the spinal cord.

Fourth Ventricle – The cavity of the medulla is the *fourth ventricle*, which communicates posteriorly with the central canal of the spinal cord. As we saw at the third ventricle, the cavity of the fourth ventricle is covered dorsally by a roof-like membrane called *tela choroidea*. A *posterior choroid plexus* extends into the fourth ventricle. Remove the tela choroidea and examine the cavity of the fourth ventricle. Note that the fourth ventricle extends into the cerebellar auricles and that they are continuous with each other beneath the cerebellum.

Motor and Sensory Columns — Columns of gray matter arising in the spinal cord may be seen in the ventral and lateral walls of the fourth ventricle. A pair of longitudinal ridges can readily be seen mid-ventrally on the floor of the ventricle. They are *somatic motor columns*. They contain the cell bodies of somatic motor neurons. Lateral to these are deep longitudinal grooves. The lateral wall of the grooves constitutes the *visceral motor columns*, containing the cell bodies of visceral motor neurons. More dorsolaterally along the walls of the fourth ventricle lie the *visceral sensory columns*, carrying impulses from visceral sensory neurons. The furthest dorsolateral wall of the cavity constitutes the *somatic sensory columns*, receiving impulses from somatic sensory neurons.

CRANIAL NERVES

The cranial nerves originate in the brain and exit at the chondrocranium. These nerves may be *sensory*, carrying impulses to the brain; they may be *motor*, carrying impulses from the brain to muscles and glands; or they may be *mixed* nerves, carrying both sensory and motor fibers.

The cranial nerves of all vertebrates have similar names and similar functions. Fish are usually described as having ten pairs of cranial nerves, the higher vertebrates have twelve.

(0) **TERMINAL NERVE** — This nerve, although found in all verebrates except cyclostomes and birds, is not numbered amongst the ten cranial nerves. Some authors consider it a part of the olfactory nerve, with which it is clearly associated. It is a very slender nerve lying along the medial surface of the *olfactory tract* and extends between the olfactory sac and olfactory lobe of the cerebrum. Its function is uncertain, some authors indicating general somatic sensory function, others visceral motor functions associated with the autonomic nervous system.

(I) **OLFACTORY NERVE** — This is a *sensory* nerve originating in the olfactory epithelium of the *olfactory sac* and terminating in the *olfactory bulb of the cerebral hemisphere. It is concerned with the sense of smell*.

(II) **OPTIC NERVE** — This is also a *sensory* nerve. It originates in the *retina* of the eye, exits the back of the orbit, passes medially and posteriorly to the *optic chiasma* and enters the *optic lobes*.

(III) **OCULOMOTOR NERVE** — This is a *motor* nerve originating from the ventral surface of the *mesencephalon*. It innervates four of the six eye muscles: the *inferior oblique*, and the *superior, inferior*, and *medial recti* muscles. It enters the orbit just posterior to the optic nerve near the superior rectus muscle. A small branch, the ciliary nerve, conveys impulses to the smooth muscles of the *iris* and ciliary body of the eye, regulating accommodation and the size of the pupil.

(IV) **TROCHLEAR NERVE** — This is a thin *motor* nerve originating in the floor of the mesencephalon, but its fibers emerge from the roof of the mesencephalon, between the optic lobe and the cerebellum. It enters the orbit of the eye between the superior rectus and superior oblique muscles via the *trochlear foramen*. It supplies the *superior oblique muscle*. Thus, five of the six eye muscles have been innervated.

Cranial nerves number V, VII, and VIII originate together by a large common stem at the anterior end of the *medulla*, just ventral to the *cerebellar auricles*.

(V) **TRIGEMINAL NERVE** — This is a *mixed* (motor and sensory) nerve which has four branches:

Superficial Ophthalmic Nerve — This *sensory* nerve passes anteriorly, medial to the orbit. It runs to the rostrum where its trigeminal fibers have a general cutaneous sensory function. It also contains facial nerve fibers arising from the common *trigeminal facial root*. Their function will be described below.

Deep Ophthalmic Nerve — This nerve passes ventral to the superficial ophthalmic nerve while delivering several small branches to the ciliary muscles within the eye. It then turns medially, makes a connection with the superficial ophthalmic trunk, and extends anteriorly supplying *sensory* fibers to the *rostrum*.

Infraorbital Nerve — A thick *sensory* nerve that runs obliquely across the floor of the orbit. Near the anterior margin of the orbit it divides into the *buccal nerve* containing fibers of the facial nerve, and a *maxillary nerve*. The maxillary nerve carries impulses back from the snout.

Mandibular Nerve — This *mixed* nerve passes laterally between the *eye* and the *otic capsule*, then turns ventrally to supply the muscles of the jaw. It is *sensory* to the muscles of the lower jaw and *motor* to the muscles of the first gill arch.

(VI) **ABDUCENS NERVE** — This *motor* nerve originates at the ventral surface of the *medulla*. It passes anterolaterally to enter the orbit where it innervates the *lateral rectus muscle* of the eye. Thus, we have accounted for the innervation of all six eye muscles.

(VII) **FACIAL NERVE** — This *mixed* nerve arises from the *medulla* and is divided into three branches:

Superficial Ophthalmic Nerve — This nerve has already been described as a branch of the trigeminal nerve. Its facial nerve fibers supply both the *ampullae of Lorenzini* and the *lateral line system*.

Buccal Nerve — This is the branch of the infraorbital nerve containing fibers of the facial nerve. It carries *sensory* impulses from the lateral line system of the head.

Hyomandibular Nerve — Near its origin this nerve exhibits a swelling, the *geniculate ganglion*. From here a small *palatine nerve* arises passing to the roof of the mouth, where it supplies the *taste buds*. The main trunk innervates the *lateral line system*, the *skin*, the *muscles of the hyoid arch*, the *tongue*, and the *floor of the mouth*.

(VIII) **AUDITORY NERVE (STATOACOUSTIC NERVE)** — This nerve arises together with cranial nerves numbers V and VII from the anterior end of the medulla. It is a short *sensory* nerve which carries impulses of the sense of equilibrium from the *inner ear*. Its two branches are the *vestibular nerve* and the *saccular nerve*.

(IX) **GLOSSOPHARYNGEAL NERVE** — This *mixed* nerve arises from the *medulla* just posterior to cranial nerve number VIII. Near the first gill pouch it exhibits a swelling called the *petrosal ganglion*. Beyond the ganglion it divides into three branches, innervating the area of the first gill arch.

Pretrematic Branch — This innervates the *first demibranch* with *sensory fibers*.

Post-trematic Branch — This branch is *mixed;* sensory to the first gill pouch and motor to the muscles of the third gill arch.

Pharyngeal Branch — This branch is *sensory* to the pharynx.

(X) **VAGUS NERVE** — This is the longest of the cranial nerves innervating gills, pharynx, esophagus, heart, stomach, intestine, and body wall. The word vagus means "wanderer." It is a *mixed* nerve which originates at the posterior end of the *medulla*. It gives off several trunks or major branches as it passes posteriorly.

Lateral Line Trunk — This sensory nerve extends beneath the *lateral line canal* posteriorly to the caudal end of the body.

Visceral Trunk — This branch gives off four branchial nerves to the remaining *gill pouches*. Each of these four *branchial nerves* divides into *pretrematic*, *post-trematic*, and *pharyngeal* branches, as did the glossopharyngeal nerve. It then continues with branches to the heart and to the anterior part of the alimentary canal. Trace the many branches of the vagus nerve.

OCCIPITAL AND SPINAL NERVES

Posteriorly, beyond the cranial nerves, two or three small occipital nerves originate. In order to observe these as well as the spinal nerves, carefully remove more muscle and cartilage from the vicinity of the spinal cord posterior to the medulla. Occipital nerves emerge from the ventral surface of the spinal cord; they have no dorsal roots. They emerge posterior to the vagus and anterior to the first spinal nerve. They join to form the *hypobranchial nerve*.

HYPOBRANCHIAL NERVE — This nerve appears to join the vagus, soon separates and is joined by the first two or three spinal nerves. The *hypobranchial nerve* carries *somatic motor* impulses to the hypobranchial muscles as well as *somatic sensory* impulses.

SPINAL NERVES — These emerge from the spinal cord between each pair of vertebra. They are formed from the union of *dorsal* and *ventral roots*. They are *mixed* nerves, innervating the skin, musculature, and viscera of the body.

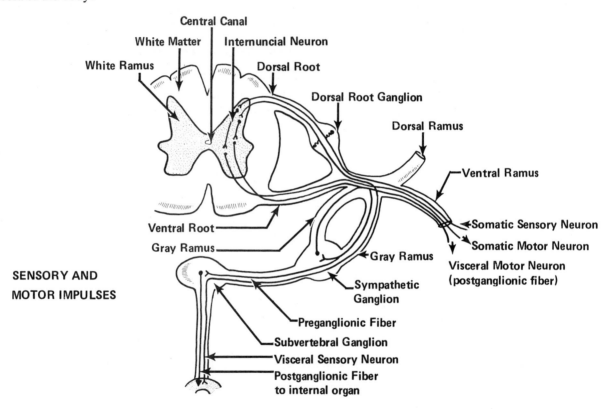

As seen in the above diagram, *sensory* and *motor* impulses pass between the body and the spinal cord.

Somatic Sensory (Afferent) Fibers — These originate as *cutaneous receptors* of temperature, pain, pressure, and touch in the *skin,* and as *proprioceptors* feeding back information about the location of the various body parts in space, the degree of stretch of skeletal muscles and tendons, and the angles between bones. They enter the spinal cord by way of the dorsal root.

Visceral Sensory (Afferent) Fibers — These bring messages from the viscera, such as pain, hunger, thirst, fullness, and nausea. They also pass to the spinal cord by way of the dorsal root.

Visceral Motor (Efferent) Fibers — These are specialized fibers passing to the organs of the *autonomic nervous system*. They include the heart, intestines, glands, uterus, and other parts of the body that function involuntarily.

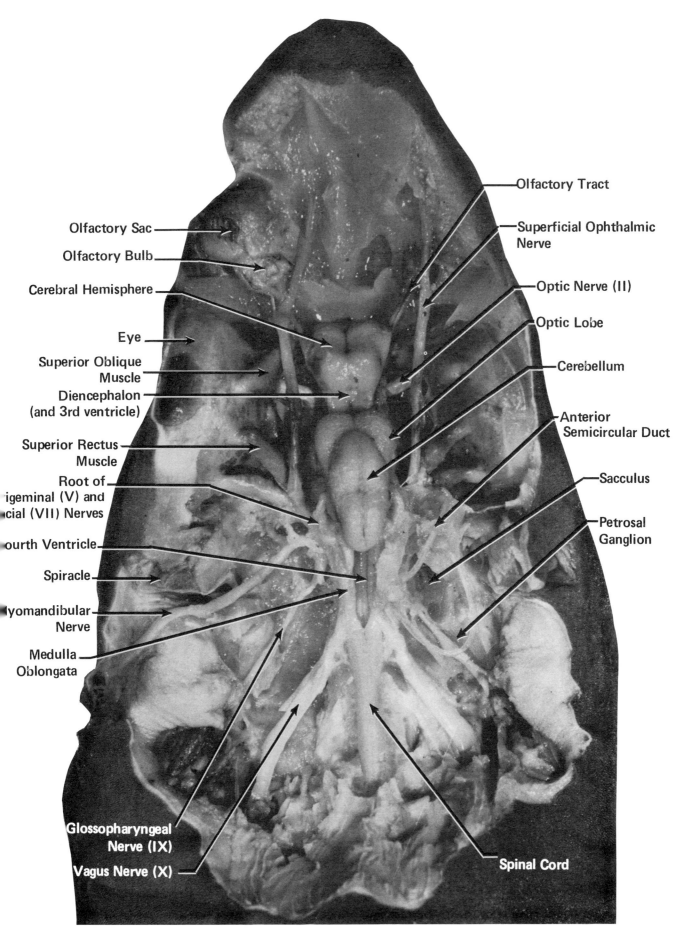

THE BRAIN AND CRANIAL NERVES – DORSAL VIEW

97

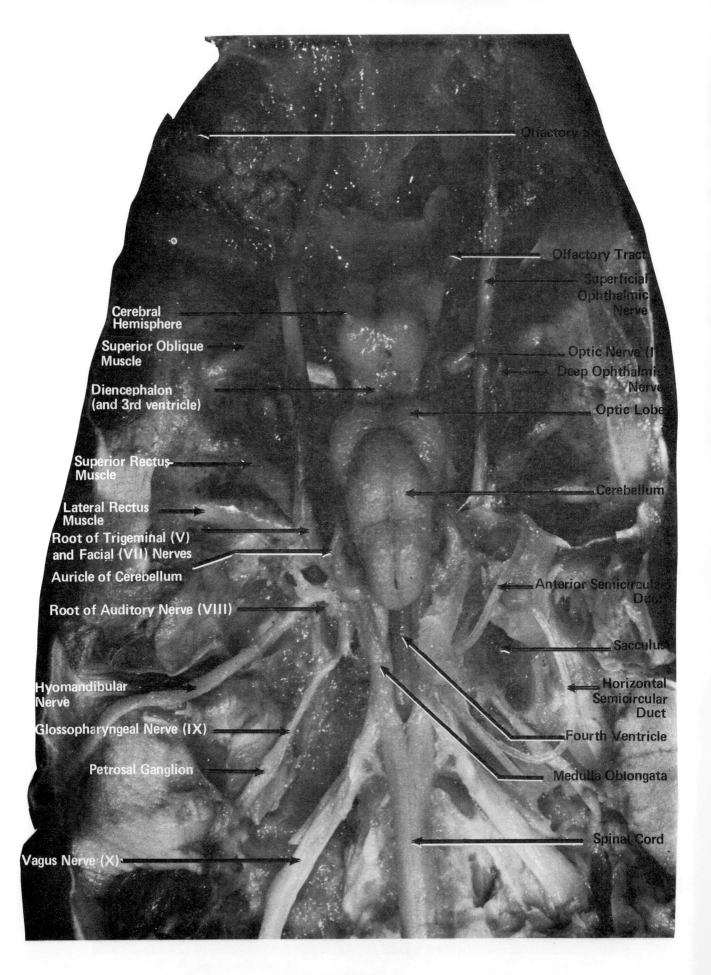

98 THE BRAIN AND CRANIAL NERVES (CLOSE-UP) — DORSAL VIEW

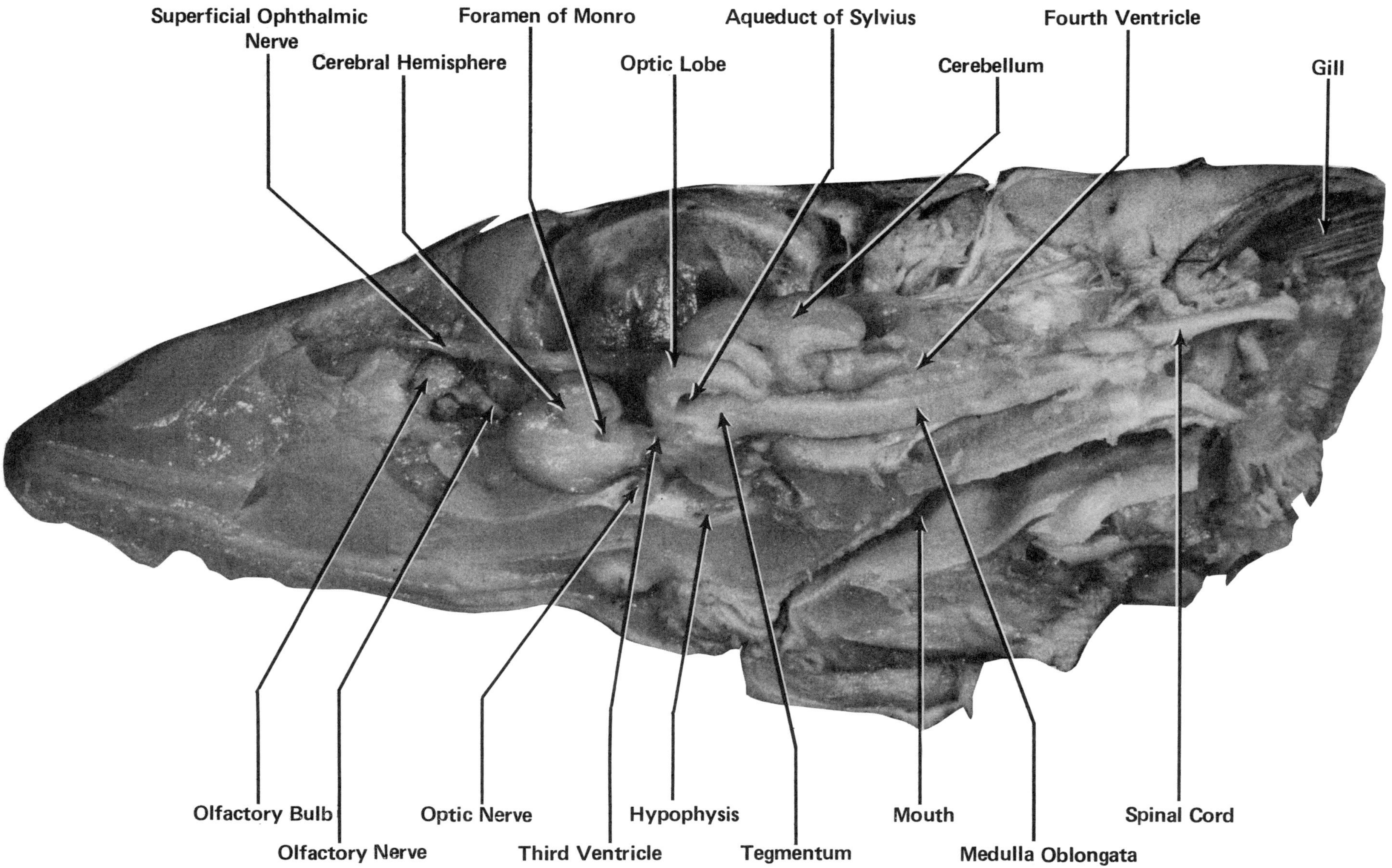

THE BRAIN – SAGITTAL SECTION

- Gill
- Fourth Ventricle
- Cerebellum
- Aqueduct of Sylvius
- Optic Lobe
- Foramen of Monro
- Cerebral Hemisphere
- Superficial Ophthalmic Nerve
- Olfactory Bulb
- Olfactory Nerve
- Optic Nerve
- Third Ventricle
- Hypophysis
- Tegmentum
- Mouth
- Medulla Oblongata
- Spinal Cord

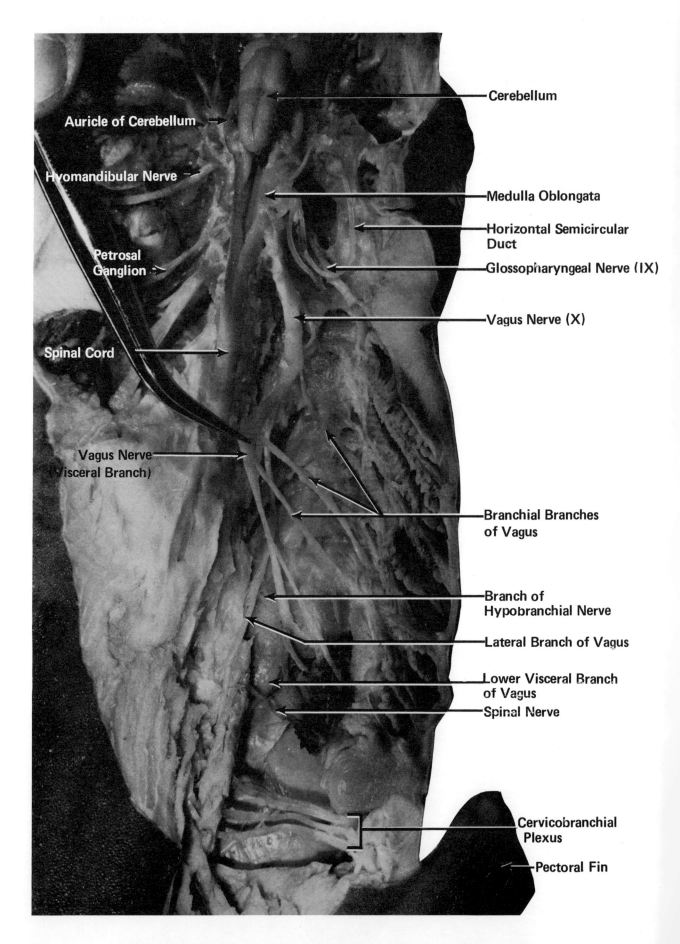

THE VAGUS AND HYPOBRANCHIAL NERVES — DORSAL VIEW

Name _____ Section _____ Date _____

SELF-QUIZ VI
THE NERVOUS SYSTEM, THE BRAIN, AND SPINAL CORD

1. Trace the nerve pathway from the nostril to the cerebral hemisphere of the brain.
2. Where are the four ventricles of the brain located?
3. What parts of the brain are included in the term diencephalon?
4. Which cranial nerves are sensory? motor? mixed?
5. What are some of the major branches of the trigeminal (V) and facial (VII) nerves?
6. To what organs does the vagus nerve (X) lead?
7. Which cranial nerves innervate the six eye muscles?
8. Describe the hypophysis and associated structures.
9. Describe the pathway of nerves to, from, and within the spinal cord.
10. Define each of the terms listed below.

ANSWERS

1. _____
2. _____
3. _____
4. _____
5. _____
6. _____
7. _____
8. _____
9. _____
10. a. cerebellum _____

    b. oculomotor nerve _____

    c. tegmentum _____

    d. pineal body _____

    e. infundibulum _____

    f. abducens _____

    g. aqueduct of Sylvius _____

    h. foramen of Monro _____

    i. habenula _____

    j. spinal ganglion _____

Label all of the features indicated on the following illustration.

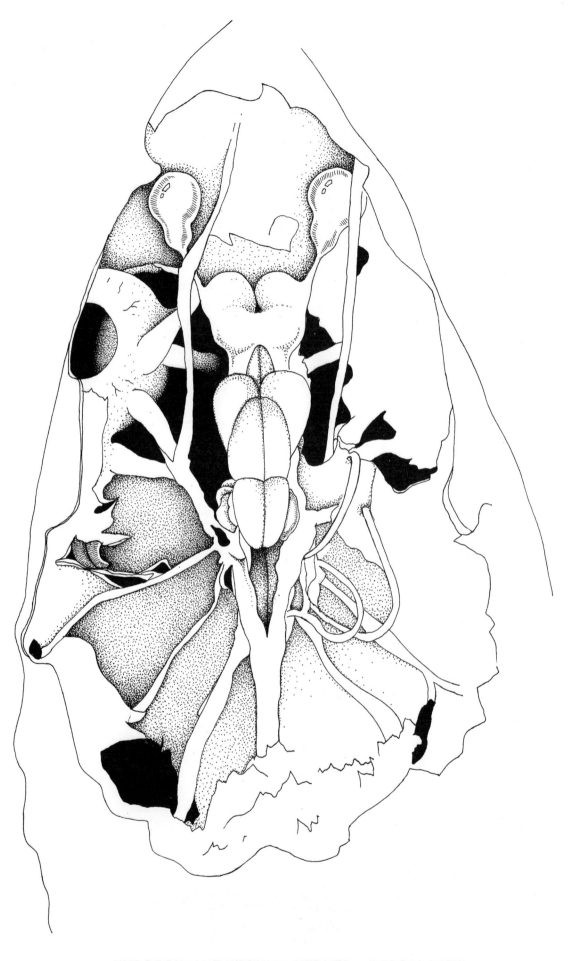

102   THE BRAIN AND CRANIAL NERVES — DORSAL VIEW

THE NERVOUS SYSTEM: THE SENSE ORGANS

A number of the shark's sense organs can be identified. Some are similar to those of higher vertebrates, others are unique to the fish. They include the *ear*, the *eye*, the *olfactory organ*, the *ampullae of Lorenzini*, the *organs of the lateral line system*, and the *pit organs*.

THE EAR

The *ear* of the shark is entirely internal. It is embedded within the *otic capsule* of the chondrocranium. Like the inner ear of higher vertebrates, its primary function is the maintaining of balance and equilibrium. There is also significant evidence that sharks detect sound waves, but its relationship to the inner ear is unclear.

The inner ear is known as the *membranous labyrinth*, and consists of a series of ducts and sacs. These canals are filled with a fluid called the *endolymph*. In sharks this is primarily sea water entering through the endolymphatic ducts. Surrounding the membranous labyrinth and following its shape is a similar series of canals in the chondrocranium, known as the *cartilaginous labyrinth,* also filled with a fluid, that protects the more delicate membranous labyrinth, the *perilymph*, which enters through the *perilymphatic ducts*. Fine connective fibers run between the cartilaginous and membranous labyrinths.

The Dissection

Locate the mid-dorsal line between the two spiracles. Look for two pores in this area. These are the openings of the *endolymphatic ducts* which lead into the *membranous labyrinth*. On the right side of the head remove the skin from the dorsal and lateral surfaces of the chondrocranium. Now, using horizontal shaving motions, remove very thin chips of chondrocranial cartilage about one millimeter thick. Look carefully if any structures can be detected below. Soon you will expose three *semicircular canals* and the sac-like structures to which they are attached.

This dissection is done on the right side of the head, since the contents of the ear have already been destroyed on the left side in exposing the brain.

SEMICIRCULAR CANALS — These are three narrow tubes about a millimeter in thickness. They are oriented in the three planes of space. One is *anterior*, the other *posterior*, and the third is *horizontal*. They are filled with *endolymph*. A disturbance of the fluid, by sudden turning or shifting of the position of the head, disturbs the endolymph.

Ampullae — Nerve endings of the cranial nerve number VIII, the *auditory nerve,* are located in a swollen area of each semicircular canal, the *ampulla*. Within the ampullae one may locate white patches of sensory epithelium called *cristae*.

Utriculi — The anterior utriculus is an enlarged area of the labyrinth that receives the anterior and horizontal semicircular canals, while the *posterior utriculus* receives the posterior semicircular canal. The utriculi also contain sensory epithelium known as the *maculae*.

SACCULUS — This is the large central chamber of the inner ear into which the endolymphatic ducts lead. The small enlargement at the posterior end of the sacculus is the *lagena*. In some fish this area has been shown to receive sound waves. White patches of sensory epithelium known as *maculae* line the inner sacculus and lagena. These patches are covered by calcareous concentrations and sand grains known as *otoliths*. Smaller otoliths are contained in the utriculi. Sudden disturbances of the endolymph and the otoliths result in sensations of imbalance. The organism responds to these by mechanisms which restore its equilibrium.

THE EYE

The eye of the shark is very similar to the eye of higher vertebrates. The one major difference is its method of accommodation in focusing for near and distant objects. While in the higher forms this is accomplished by changing the shape of the lens, in the shark it is done as in most cameras by moving the lens further or closer to the retina.

The Dissection

The dissection will be done in two stages. First, the eyeball will be removed from its socket (the orbit) in order to view some of the muscles, nerves, and supporting structures of the orbit and the external parts of the eyeball. Then we shall cut into the eye to view its internal structure.

EYELIDS — Although most fish have no *eyelids*, the shark does possess eyelid folds. However, unlike the eyelids of the higher vertebrates, those of the shark are immovable.

CONJUNCTIVA — A thin transparent membrane which covers the outer surface of the eye. It folds at its outer edge and passes under the lid.

The Dissection

Make a circular cut around the edge of the eye at the junction of the conjunctiva and eyelid to separate the eyeball from its orbit. Remove some of the dorsal chondrocranium over the eye. You will expose the six extrinsic eye muscles and other structures to be described below. Cut the eye muscles and nerves near their insertions on the eyeball. Lift out and remove the eyeball from its orbit.

EYE MUSCLES — Of the six extrinsic eye muscles, two originate in the anteromedial orbital wall; these are the *oblique muscles* which pull the eye diagonally. The other four originate in the posteromedial wall of the orbital wall; these are the *rectus* muscles which pull straight back on the eyeball. The nerves which innervate these muscles were described in the last chapter, see page 94.

Oblique Muscles (Superior and Inferior) — As the names indicate, one of the oblique muscles is inserted superiorly, or on the dorsal portion of the orbit; the second inferiorly, or on the ventral surface.

Rectus Muscles (Superior, Inferior, Medial, and Lateral) — The names of these muscles indicate the directions in which the eye is moved. The *superior rectus* which moves the eye upwards is inserted on the dorsal surface. The *inferior rectus* which moves the eye downwards is inserted on the ventral surface. The *medial rectus* which moves the eye anteriorly is inserted on the anterior surface of the eye. The *lateral rectus* which moves the eye posteriorly is inserted on the posterior surface of the eye.

OPTIC PEDICEL — This cartilaginous structure, shaped like a golf tee, projects from near the origin of the rectus muscles. Its flattened disc-like distal end rests against and supports the eyeball medially and aids in its rotation.

OPTIC NERVE — A thick white stump, the *optic nerve* may be seen exiting the back of the eyeball. It emerges just ventral to the site of the insertion of the medial rectus muscle.

OTHER NERVES OF THE EYE — Nerves previously described in the discussion of cranial nerves may also be found in the orbit. These include the *deep opthalmic nerve, oculomotor nerve, trachlear nerve, infraorbital nerve, abducens nerve,* and the *superficial ophthalmic nerve.*

The Dissection

The eyeball has already been removed from its orbit. We are now ready to observe the eyeball itself, both externally and internally. First observe the external features, then cut through the eyeball about halfway between dorsal and ventral surfaces. This will expose the inner structures. See photos on pages 110 and 112.

SCLERA — This is the tough white fibrous outer coat of the eye. At places it is made even more firm by cartilage embedded in the sclera.

CORNEA — At the front of the eye this tough coat becomes transparent as the cornea of the eye. The *conjunctiva* lies over the cornea.

CHOROID — This is the vascular black pigmented middle layer of the eye. Laterally, it is fused to the *retina*. The darkly pigmented layers help in absorbing light within the eye.

SUPRACHOROIDEA — A thick non-pigmented layer between the *sclera* and *choroid*. It is composed of connective tissue, lymph spaces and vascular tissue. It is only found in species which possess an *optic pedicel*.

RETINA — This is the multi-layered sensory gray-white colored membrane. The *rods* and *cones* which receive light stimuli are located here. The optic nerve leaving the eye is a continuation of the light receptor cells in this membrane.

PUPIL — This round hole behind the cornea is an opening in the *iris* of the eye. It can dilate or constrict to allow more or less light to enter.

IRIS — A pigmented anterior extension of the *choroid* layer. In its center is the *pupil*. The iris regulates the size of the pupil.

CILIARY BODY — This is another extension of the black pigmented *choroid*. It is a thin black band indented with faint radial lines. It lies between the iris and the choroid.

LENS — While in preserved specimens the spherical *lens* is opaque and hard, in living organisms it is clear and flexible. It lies behind the *pupil*. It helps to focus the light upon the light sensitive retina. It is suspended in the eyeball by a *suspensory ligament* which originates from the *ciliary body* and is attached to the equator of the lens.

VITREOUS CHAMBER — The main cavity of the eyeball contains a gelatinous, transparent semi-solid called the *vitreous humor*. It gives shape to the eyeball and prevents it from collapsing.

ANTERIOR CHAMBER — This chamber, much smaller than the vitreous chamber, lies anteriorly between the cornea and the iris. It contains a clear watery fluid, the *aqueous humor*.

OLFACTORY ORGAN

This is the shark's organ of smell. Note the location of the external *nares* (nostrils). Each is divided into two openings: the lateral one, an *incurrent aperture,* and the medial one, an *excurrent aperture*. These are partially separated by a flap of skin that regulates the flow of water into and out of the nares, as in the photo on page 113. There is no connection between the nasal area and the mouth or oral cavity.

The Dissection

Make a transverse cut across the snout through the center of one of the nares. Observe the following:

OLFACTORY SACS — These bulb-like structures, spherical in shape, contain a series of radial folds which increase the surface area called *olfactory lamellae*. Their surfaces are covered with *olfactory epithelium*. Sea water taken into the nares is passed over these sensory areas. Here the odors stimulate the cilia-like endings of neuro-sensory cells.

OLFACTORY BULBS — These are a paired anterior extension of the brain leading into the posterior end of the olfactory sacs. Their fibers continue into the *olfactory tract* and the *olfactory lobe* of the *cerebral hemisphere*.

LATERAL LINE SYSTEM

This sensory system is only found in fish and amphibian larvae.

LATERAL LINE CANALS — It is a system of sensory cells and canals under the skin which responds to mechanical movement of the water, to changes in water pressure and other disturbances in the water. It consists of a series of interconnected *lateral line canals*. One long lateral line canal runs the length of the body on either side along the lateral surface. This can be readily observed as a thin light-colored line upon the skin. Other canals are found in the head region, near the eyes and jaws. The canals open to the outside of the skin by tiny *pores* which admit water.

The interconnected lateral line canal system in the region of the head includes the *infraorbital canal,* which passes ventrally posterior to the eye, then extends to the tip of the snout. The *supraorbital canal* passes forward above the eye to the snout, then turns to extend posteriorly to connect with the infraorbital canal. A *supratemporal canal* passes over the top of the head at the level of the spiracles. Several smaller canals may also be detected in the areas of the jaws. These are all a part of the *lateral line system*.

Neuromasts — These are the ciliated sensory cells lining the canals that can detect water currents.

The Dissection

Remove a section of skin, about two inches by two, along the lateral surface of the body. Examine the muscular body wall to detect the *lateral line canal*. Use a hand lens to find some of the pores along the lateral line on the surface of the skin.

AMPULLAE OF LORENZINI

These sense organs are modifications of the lateral line system, and are similarly innervated. They detect changes in water temperature, electric current, and salinity.

Examine the underside of the snout. Note many large *pores*. Press down upon these areas and observe the gelatinous secretion exuded. The pores connect to elongated, cylindrical, tube-like canals which store the jelly-like secretion. At their bases they form the main bulb-like *ampullae of Lorenzini*. Several secondary ampullae may also be seen. At the base of the ampullae find the sensory nerve by which it is innervated.

The Dissection

Remove the skin from the underside of the snout and observe the structures just described.

PIT ORGANS

These sensory areas are also innervated by *neuromast-like cells*. Their pores may be found in a row near the bases of the pectoral fins and anterior to the first gill slit. They are believed to aid the shark in the detection of temperature changes.

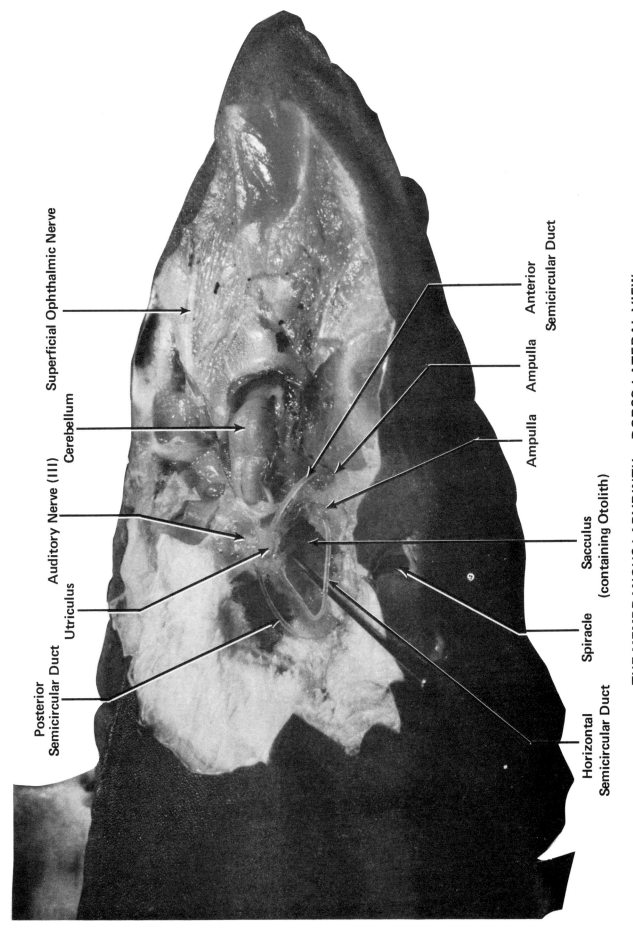

THE MEMBRANOUS LABYRINTH – DORSO-LATERAL VIEW

THE EYE – DORSO-LATERAL VIEW

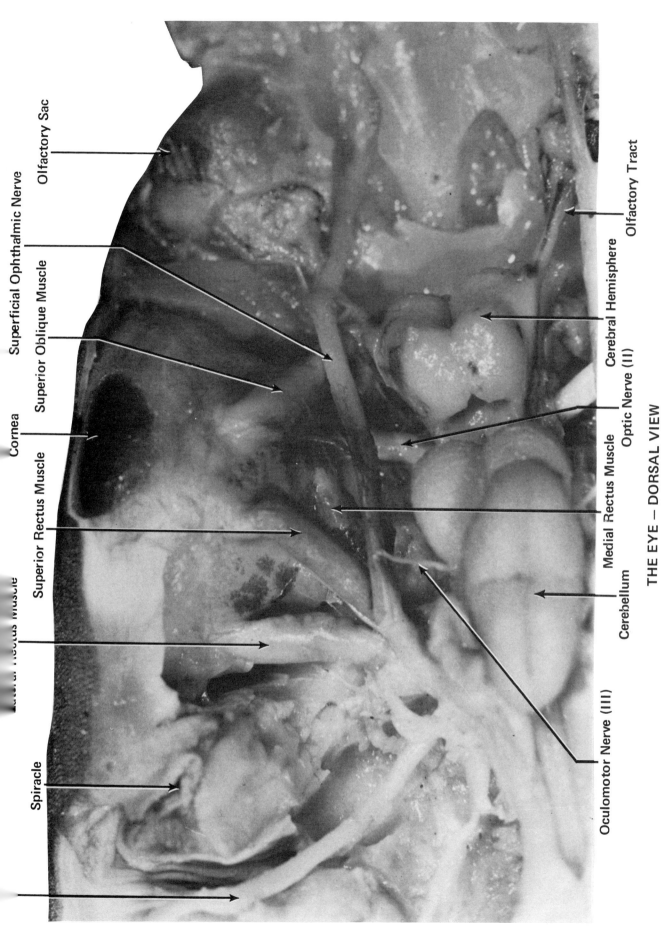

THE EYE – DORSAL VIEW

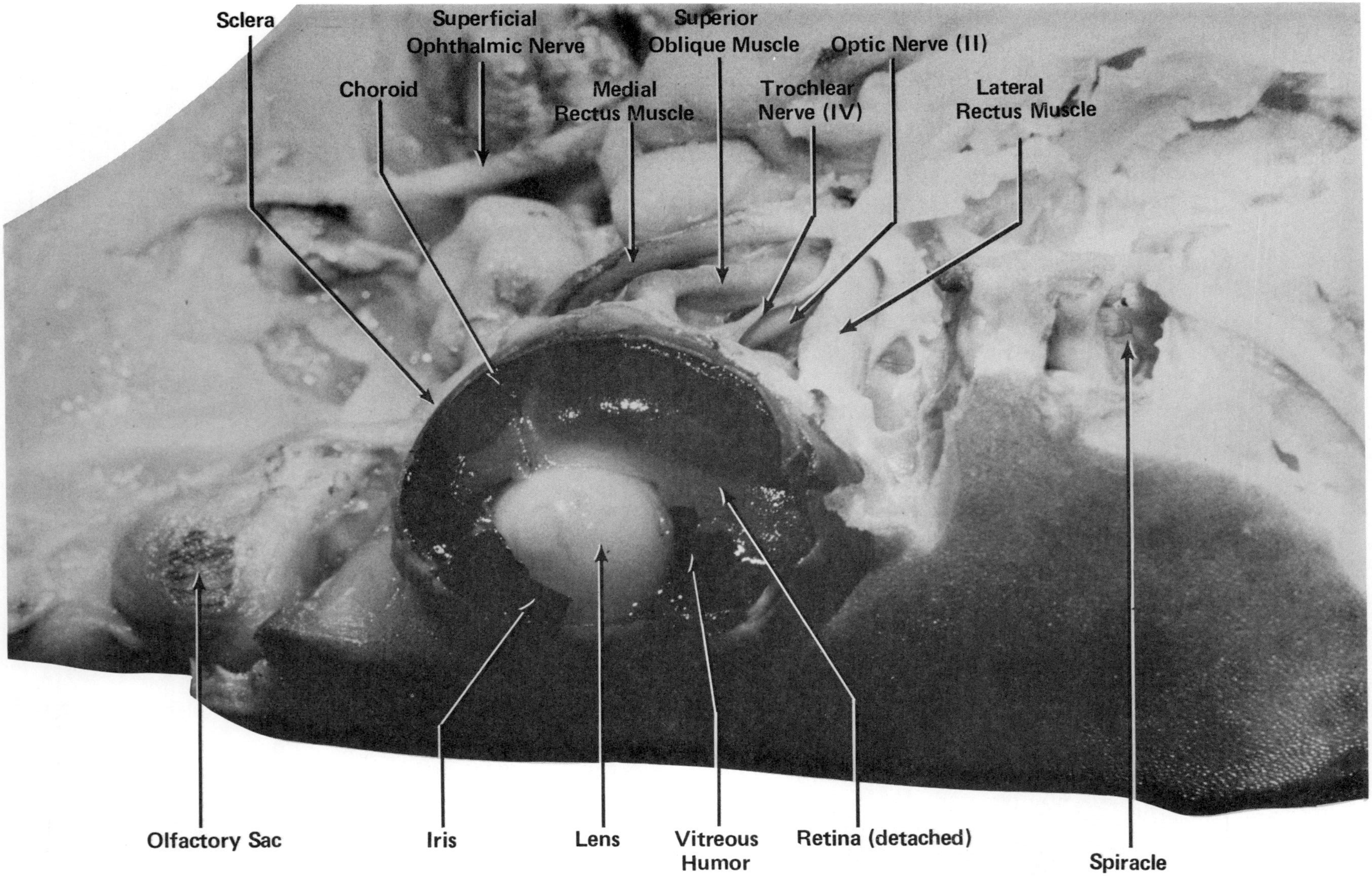

Sclera

Superficial
Ophthalmic Nerve

Superior
Oblique Muscle   Optic Nerve (II)

Trochlear
Nerve (IV)   Lateral
Rectus Muscle

Medial
Rectus Muscle

Choroid

Spiracle

Retina (detached)

Vitreous
Humor

Lens

Iris

Olfactory Sac

THE EYE – SAGITTAL SECTION

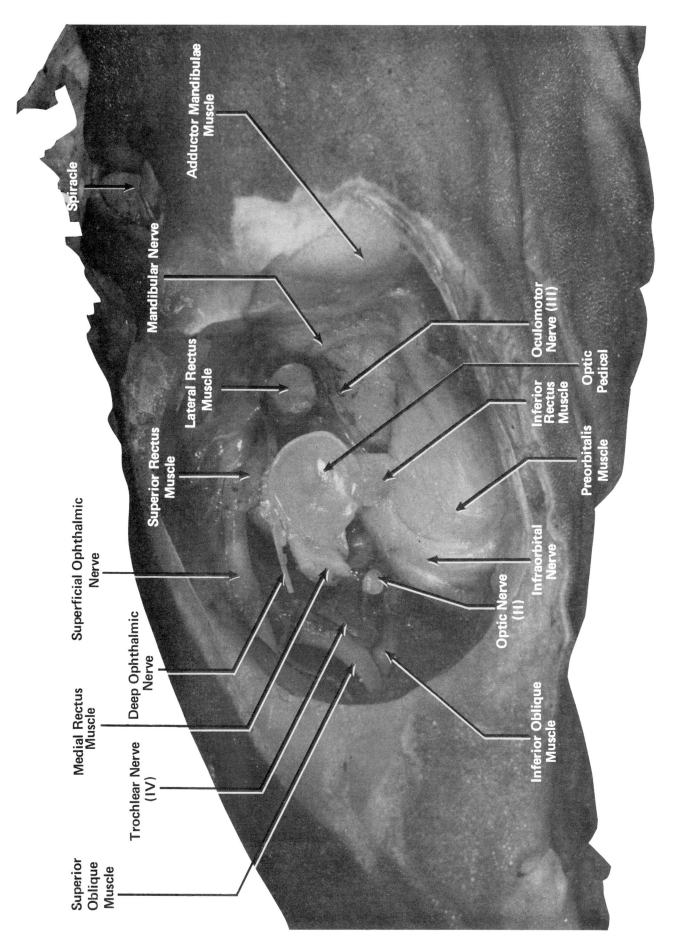

THE LEFT ORBIT – LATERAL VIEW

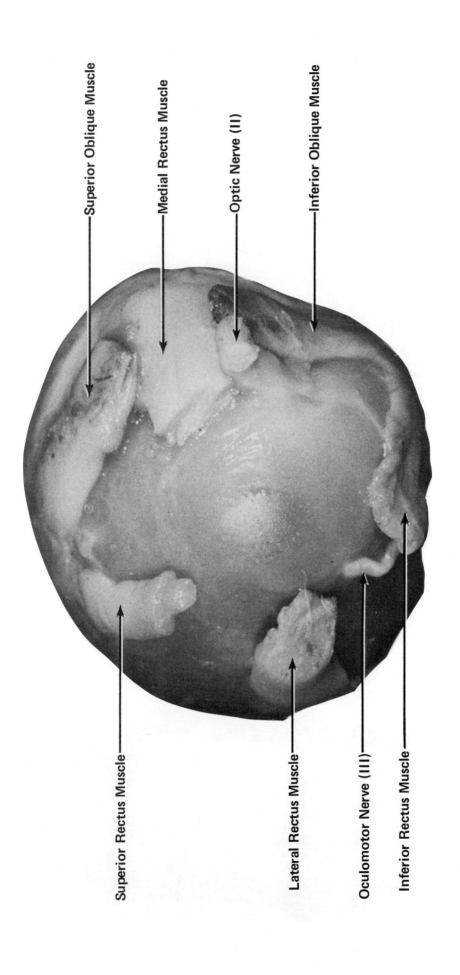

THE LEFT EYE — MEDIAL VIEW

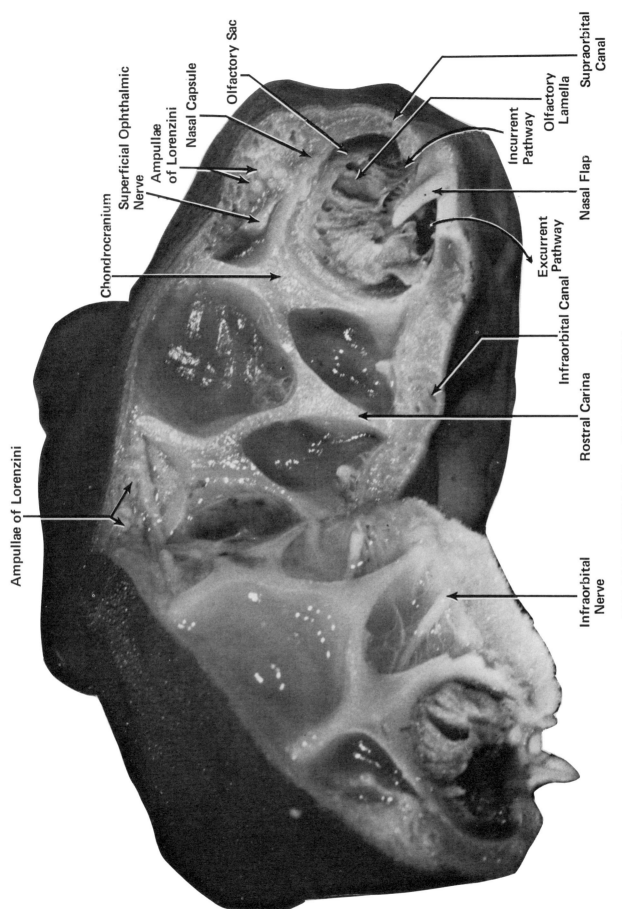

THE OLFACTORY SAC — CROSS SECTION

Name _____ Section _____ Date _____

SELF-QUIZ VII
THE NERVOUS SYSTEM: THE SENSE ORGANS

1. What are the functions of the membranous labyrinth?
2. Name five parts of the eye and their functions.
3. What are the functions of the ampullae of Lorenzini?
4. Describe the lateral line system of the trunk.
5. Describe the lateral line system of the head.
6. Describe the orbit of the eye after the eyeball has been removed.
7. How does the eye of the shark differ from the eye of man?
8. Describe the inner structure of the olfactory sac.
9. Name the six extrinsic eye muscles and the directions in which they move the eye.
10. Define each of the terms listed below.

ANSWERS

1. _____
2. _____
3. _____
4. _____
5. _____
6. _____
7. _____
8. _____
9. _____
10. a. optic pedicel _____
    b. olfactory lamella _____
    c. pit organ _____
    d. utriculus _____
    e. otolith _____
    f. sacculus _____
    g. endolymphatic duct _____
    h. choroid _____
    i. ciliary body _____
    j. sclera _____

Label all of the features indicated on the following illustration.

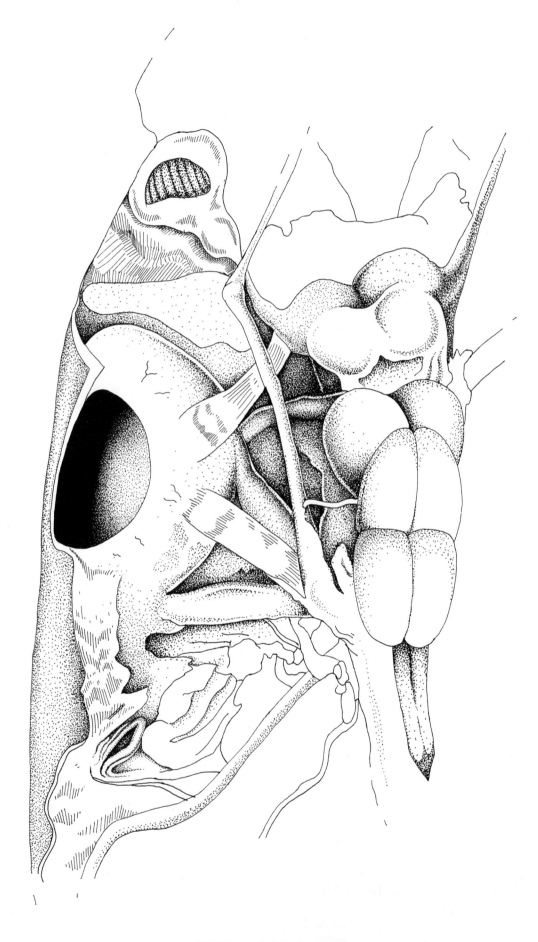

THE EYE — DORSAL VIEW

SHARK DISSECTION CHECK LIST

EXTERNAL FEATURES
___ **The Skin**
  ___ placoid scales
  ___ lateral line canal
  ___ ampullae of Lorenzini
___ **The Head**
  ___ rostrum
  ___ nares
  ___ jaws
    ___ teeth
  ___ eyes
    ___ cornea
    ___ pupil
    ___ conjunctiva
  ___ spiracles
  ___ gill slits
___ **The Fins**
  ___ dorsal fins
    ___ spines
  ___ caudal fin
  ___ pectoral fins
  ___ pelvic fins
___ **The Cloaca**
  ___ claspers

THE SKELETAL SYSTEM
___ **The Axial Skeleton**
  ___ the skull — chondrocranium
    ___ neurocranium
      ___ rostrum
      ___ nasal capsules
      ___ orbits
      ___ otic capsules
      ___ foramina
      ___ notochord
      ___ occipital region
    ___ splanchnocranium
      ___ visceral arches
        ___ mandibular arch
          ___ palatoquadrate
          ___ Meckel's cartilage
        ___ hyoid arch
          ___ basihyal
          ___ ceratohyal
          ___ hyomandibular
        ___ gill arches
          ___ pharyngobranchial
          ___ epibranchial
          ___ ceratobranchial
          ___ hypobranchial
          ___ basibranchial
          ___ gill rakers
          ___ gill rays
  ___ the vertebral column
    ___ tail vertebrae
      ___ centrum
      ___ neural arch
      ___ spinal cord
      ___ vertebral canal
      ___ hemal arch
      ___ hemal canal
      ___ neural spine
      ___ hemal spine
    ___ trunk vertebrae
      ___ basapophyses
___ **The Appendicular Skeleton**
  ___ pectoral girdle
    ___ coracoid bar
    ___ scapular cartilage
    ___ suprascapular cartilage
  ___ pectoral fin
    ___ basal cartilages
      ___ propterygium
      ___ mesopterygium
      ___ metapterygium
    ___ radial cartilages
    ___ ceratotrichia
  ___ pelvic girdle
    ___ puboischiac bar
      ___ iliac process
  ___ pelvic fin
    ___ basal cartilages
      ___ propterygium
      ___ metapterygium
    ___ radial cartilages
    ___ ceratotrichia
  ___ dorsal fin
  ___ caudal fin

THE MUSCULAR SYSTEM
___ myotomes
___ myosepta
  ___ horizontal septum
  ___ dorsal median septum
  ___ ventral median septum

___ epaxial muscles
___ hypaxial muscles
___ linea alba
___ head and branchial regions
    ___ superficial constrictors
        ___ first dorsal constrictor
            ___ adductor mandibulae
            ___ spiricularis
            ___ preorbitalis
        ___ first ventral constrictors
            ___ intermandibularis
        ___ second constrictor
            ___ hyoid constrictor
            ___ interhyoid
        ___ third to sixth constrictors
            ___ dorsal constrictors
            ___ ventral constrictors
    ___ levators
        ___ first levator
            ___ palatoquadrati
            ___ hyomandibulae
        ___ third to sixth levators
            ___ cucullaris
    ___ interarcuals
        ___ dorsal interarcuals
        ___ lateral interarcuals
        ___ subspinal
        ___ branchial adductors
    ___ hypobranchial
        ___ common coracoarcuals
        ___ coracohyoids
        ___ coracomandibular
        ___ coracobranchials
    ___ epibranchial
___ appendicular muscles
    ___ fin musculature
        ___ pectoral fins
            ___ flexors and extensors
        ___ pelvic fins
            ___ flexors and extensors
        ___ dorsal fins
            ___ radial muscles

THE DIGESTIVE AND RESPIRATORY SYSTEMS
___ **Pleuroperitoneal Cavity**
    ___ coelom
    ___ transverse septum
    ___ peritoneum
    ___ liver
        ___ 3 lobes
    ___ gall bladder
        ___ common bile duct
    ___ falciform ligament
    ___ coronary ligament
    ___ esophagus
    ___ stomach
        ___ greater curvature
        ___ lesser curvature
        ___ mesogaster
        ___ lesser omentum
        ___ rugae
        ___ pyloric sphincter
    ___ duodenum
    ___ pancreas
    ___ spleen
    ___ valvular intestine
        ___ spiral valve
    ___ rectal gland
    ___ cloaca
    ___ abdominal pores
___ **Oral Cavity and Pharynx**
    ___ teeth
    ___ tongue
    ___ spiracles
    ___ gills
        ___ gill lamellae
        ___ internal gill slits
        ___ gill arches
        ___ gill rakers
        ___ branchial bars
        ___ branchial blood vessels
        ___ interbranchial septa
        ___ gill arch cartilages
        ___ gill rays
        ___ valve flaps
        ___ demibranchs
        ___ holobranchs
        ___ pseudobranchs

THE CIRCULATORY SYSTEM
___ **Pericardial Cavity**
    ___ pericardium
___ **The Heart**
    ___ sinus venosus
    ___ atrium
    ___ ventricle
    ___ conus arteriosus
    ___ sinoatrial aperture
    ___ arterioventricular aperture
___ **Ventral Aorta and Afferent Branchial Arteries**

___ **Efferent Branchial Arteries**
    ___ efferent collector loops
        ___ pretrematic branch
        ___ post-trematic branch
        ___ intertrematic branches
    ___ dorsal aorta
    ___ external carotid artery
    ___ hyoidean epibranchial artery
        ___ stapedial artery
        ___ internal carotid artery
        ___ efferent spiricular artery
    ___ afferent spiricular artery
    ___ ophthalmic artery
    ___ paired dorsal aortae
    ___ commissural artery
    ___ pericardial artery
    ___ coronary artery
    ___ esophageal artery
___ **Branches of Dorsal Aorta**
    ___ subclavian artery
    ___ lateral artery
    ___ ventrolateral artery
    ___ brachial artery
    ___ celiac artery
        ___ gastrohepatic artery
        ___ gastric artery
        ___ hepatic artery
        ___ pancreaticomesenteric artery
    ___ anterior mesenteric artery
    ___ lienogastric artery
    ___ posterior mesenteric artery
    ___ renal arteries
    ___ iliac arteries
    ___ caudal artery
___ **The Venous System**
    ___ hepatic portal vein
        ___ gastric vein
        ___ lienomesenteric vein
            ___ posterior intestinal vein
            ___ posterior lienogastric vein
        ___ pancreaticomesenteric vein
            ___ intraintestinal vein
            ___ anterior lienogastric vein
            ___ anterior intestinal vein
    ___ caudal vein
    ___ renal portal veins
        ___ afferent renal veins
        ___ efferent renal veins
___ **The Systemic Veins**
    ___ hepatic veins
    ___ common cardinal veins
        ___ anterior cardinal veins
            ___ anterior cardinal sinus
            ___ orbital sinuses
        ___ inferior jugular veins
            ___ hyoidean sinus
        ___ posterior cardinal veins
            ___ genital sinuses
    ___ subclavian veins
        ___ brachial veins
        ___ subscapular veins
        ___ lateral abdominal veins
            ___ cloacal veins
            ___ femoral veins
            ___ iliac veins
___ coronary veins

THE UROGENITAL SYSTEM
___ kidneys
___ archinephric duct
___ urinary papilla
___ female genital system
    ___ ovaries
        ___ mesovarium
    ___ oviducts
        ___ mesotubarium
        ___ ostium
        ___ shell gland
    ___ uterus
    ___ cloaca
___ male genital system
    ___ testes
        ___ mesorchium
    ___ epididymis
    ___ ductus deferens
    ___ seminal vesicle
    ___ sperm sacs
    ___ accessory urinary ducts
    ___ urogenital sinus
    ___ urogenital papilla
    ___ cloaca
    ___ siphon
    ___ claspers
        ___ clasper tube

THE NERVOUS SYSTEM
___ **The Brain**
    ___ primitive meninx
    ___ prosencephalon
        ___ telencephalon
            ___ olfactory bulbs

  ____ olfactory tracts
  ____ cerebral hemispheres
  ____ olfactory lobes
 ____ diencephalon
  ____ third ventricle
   ____ foramen of Monro
   ____ aqueduct of Sylvius
  ____ epithalamus
   ____ tela choroidea
   ____ epiphysis
  ____ thalamus
  ____ hypothalamus
____ mesencephalon
 ____ optic lobes
____ rhombencephalon
 ____ metencephalon
  ____ cerebellum
 ____ myelencephalon
  ____ medulla oblongata
  ____ fourth ventricle
   ____ motor and sensory columns

____ **Cranial Nerves**
 ____ terminal nerve
 ____ olfactory nerve
 ____ optic nerve
 ____ oculomotor nerve
 ____ trochlear nerve
 ____ trigeminal nerve
  ____ superficial ophthalmic nerve
  ____ deep ophthalmic nerve
  ____ infraorbital nerve
  ____ mandibular nerve
 ____ abducens nerve
 ____ facial nerve
  ____ buccal nerve
  ____ hyomandibular nerve
   ____ geniculate ganglion
 ____ auditory nerve
  ____ vestibular nerve
  ____ saccular nerve
 ____ glossopharyngeal nerve
  ____ pretrematic branch
  ____ post-trematic branch
  ____ pharyngeal branch
 ____ vagus nerve
  ____ lateral line trunk
  ____ visceral trunk

____ **Occipital and Spinal Nerves**
 ____ hypobranchial nerve
 ____ spinal nerves

____ **The Sense Organs**
 ____ the ear
  ____ membranous labyrinth
  ____ cartilaginous labyrinth
  ____ perilymphatic ducts
  ____ semicircular canals
   ____ anterior, posterior, horizontal
   ____ ampullae
   ____ utriculi
   ____ sacculus
    ____ lagena
    ____ maculae
    ____ otoliths
 ____ the eye
  ____ eyelids
  ____ conjunctiva
  ____ eye muscles
   ____ oblique muscles (2)
   ____ rectus muscles (4)
  ____ optic pedicel
  ____ optic nerve
  ____ sclera
  ____ cornea
  ____ choroid
  ____ suprachoidea
  ____ retina
  ____ pupil
  ____ iris
  ____ ciliary body
  ____ lens
   ____ suspensory ligament
  ____ vitreous chamber
  ____ anterior chamber
 ____ olfactory organ
  ____ nares
   ____ incurrent aperture
   ____ excurrent aperture
  ____ olfactory sacs
   ____ olfactory lamellae
  ____ olfactory bulbs
 ____ lateral line system
  ____ lateral line canals
   ____ infraorbital canal
   ____ supraorbital canal
   ____ supratemporal canal
  ____ neuromasts
 ____ ampullae of Lorenzini
 ____ pit organs